CAR MAINTENANCE AND REPAIR BIBLE

Master DIY Car Fixes to Avoid Breakdowns & Costly Bills, Diagnose Common Issues & Save Thousands on Mechanics | Insider's Tips for At-Home Car Maintenance

Bill J. Paxson

TABLE OF CONTENTS

INTRODUCTION

WELCOME TO THE SECRET BIBLE: THE KNOWLEDGE MECHANICS DON'T WANT YOU TO HAVE

Let me ask you something: have you ever walked out of a mechanic's shop with a receipt that made your stomach drop? Maybe it was for something like an oil change or replacing brake pads—things you later realized aren't rocket science. If that sounds familiar, you're not alone. For most people, the world under the hood is a mystery. All those parts, tools, and technical terms make it seem like car maintenance is best left to the professionals.

But here's the truth: most of those expensive repairs aren't as complicated as they're made out to be. Mechanics know this, and while they provide a necessary service, the industry benefits from what car owners don't know. The less you understand about your vehicle, the more you rely on someone else to fix it—and that reliance comes with a hefty price tag.

That's where this book comes in. Think of it as your personal guide to mastering car care. It's not a dense, jargon-filled manual; it's a straightforward roadmap designed to empower you. By the time you're done reading, you'll know how to take control of your car's maintenance, save hundreds (if not thousands) of dollars, and feel confident handling issues when they arise.

Whether you're here because you're tired of overpriced repairs or you just want to know more about how your car works, you're in the right place. Let's dive into how learning even the basics of car care can transform your experience as a driver.

WHY I WROTE THIS BOOK

Owning a car is liberating, but it's also a responsibility. Keeping your car in good shape doesn't just protect your investment—it ensures your safety and saves you money. But, after decades in the automotive industry, I've seen countless car owners pay far too much for basic services. Simple repairs like replacing a brake pad or fixing a tire puncture are billed as premium services at repair shops. Why? Because most people don't know these tasks are well within their capabilities.

I wrote this book to share the secrets mechanics often keep to themselves. My goal is to empower everyday drivers with the skills and confidence to take charge of their car's care. This book is my way of leveling the playing field so you no longer feel at the mercy of your mechanic.

WHO AM I, AND WHY SHOULD YOU TRUST ME?

My name is Bill J. Paxson, and I've been a certified mechanic for several years. Cars have been my passion since I was a teenager, tinkering in my parents' garage. Since then, I've worked in repair shops, taught workshops, and helped countless car owners navigate the intimidating world of car care.

What sets me apart is my ability to break down complex automotive concepts into simple, actionable steps. Whether you're a complete beginner or someone looking to refine their DIY skills, I've tailored this book to meet you where you are.

WHY THIS BOOK WILL SAVE YOU THOUSANDS IN REPAIRS

Owning a car comes with its fair share of expenses. There's the cost of the car itself, insurance, gas, and those unexpected repair bills that always seem to come at the worst time. While some expenses are unavoidable, repair bills are often an area where you can take back control—and save a lot of money.

Think about it: how much do you spend each year on repairs and maintenance? For many car owners, it's anywhere from a few hundred to a few thousand dollars. The frustrating part? A significant chunk of that money goes toward tasks you could easily do yourself.

Let's break it down. Take something as simple as an oil change. At a mechanic's shop, you'll pay between $60 and $100 for the service. If you do it yourself, you're looking at $20 to $30 for the oil and filter and about 30 minutes of your time. Over a decade of driving, those savings add up fast.

Then there are other routine tasks—like changing an air filter, rotating tires, or replacing windshield wiper blades. These jobs aren't just simple; they're also some of the most overcharged services at repair shops. For example, a mechanic might charge $50 to $70 to replace an air filter, but buying and installing one yourself costs less than $20.

Here's where things get even more important: skipping maintenance doesn't just hurt your wallet in the short term; it can lead to bigger, costlier problems down the road. Neglecting basic tasks like checking coolant levels, replacing worn brake pads, or inspecting your tires can result in serious damage. A blown radiator or failed brakes can easily cost thousands of dollars—and that's if you're lucky enough to avoid an accident.

This book is about helping you take a proactive approach. By learning the basics of car care and establishing good maintenance habits, you'll keep your vehicle running smoothly and avoid those dreaded surprise breakdowns.

HOW TO USE THIS BOOK TO MAINTAIN AND REPAIR YOUR CAR

This isn't just a book filled with tips—it's a hands-on guide that takes you from feeling clueless about car maintenance to handling it like a pro. Whether you're a total beginner or someone who's dabbled in DIY car repairs, this book is here to meet you where you are. Here's what you can expect:

We'll begin with the basics. The first chapters explain how your car's major systems work—like the engine, brakes, and electrical components. You'll learn what these systems do, how to spot early warning signs of trouble, and why regular maintenance matters. Don't worry about feeling overwhelmed; this section is designed to be approachable and easy to understand, even if you've never looked under the hood before.

You don't need a fully stocked garage to take care of your car. With just a few affordable tools, you can handle most maintenance and repair tasks at home. This book includes a simple guide to building your DIY toolkit, with recommendations for everything from basic wrenches to diagnostic devices like an OBD-II scanner. We'll also share tips on setting up a functional workspace, whether you're working in a garage or a driveway.

Preventative care is the cornerstone of this book. You'll learn how to establish a regular routine for checking fluids, inspecting tires, and keeping your battery in good shape. These small, consistent habits can save you from expensive repairs and give you peace of mind every time you drive.

When something goes wrong, don't panic. The middle chapters of this book focus on detailed, step-by-step instructions for common repairs. Whether it's changing your brake pads, replacing spark plugs, or diagnosing a dashboard warning light, you'll find clear guides that walk you through the process. These instructions are written with beginners in mind, so you won't feel out of your depth.

As much as this book empowers you to tackle car care, it's also important to know when to call in a professional. Some repairs require specialized tools or experience, and that's okay. We'll show you how to recognize those situations, find a trustworthy mechanic, and avoid being overcharged for services.

The more you learn about your car, the more confident you'll feel. By the time you finish this book, you'll understand your vehicle inside and out. You'll know what's normal, what needs attention, and how to handle problems without feeling overwhelmed.

Let's face it—car maintenance can be intimidating, especially when you're starting from scratch. The first time you pop your hood and look at the maze of parts and wires, it's easy to feel overwhelmed. But here's the thing: you don't need to figure everything out all at once. Like any new skill, car care is something you can learn step by step.

A lot of people worry about making mistakes. What if you accidentally damage something? What if you don't have the right tools? Those concerns are valid, but they're also manageable. This book is designed to guide you through those fears with clear instructions and plenty of encouragement. You'll start small—think checking tire pressure or replacing a wiper blade—and gradually work your way up to more complex tasks.

Most of us think of our cars as tools—something to get us from point A to point B. But when you start taking care of your car yourself, it becomes something more. You'll develop a deeper understanding of how it works and feel a sense of pride in keeping it running. Every repair or maintenance task you complete will build your confidence and strengthen your connection to your vehicle.

The financial benefits of DIY car care are obvious. Over time, the money you save on repairs and maintenance can add up to thousands of dollars. But the benefits go beyond your wallet. When you know how to care for your car, you'll also save time, reduce stress, and avoid the frustration of unexpected breakdowns. You'll gain the independence that comes with knowing you can handle whatever comes your way.

You don't have to be an expert to take control of your car's care. You just need the right guidance—and that's what this book is here to provide. By the end of this journey, you'll have gone from feeling unsure and dependent to confident and capable. You'll save money, gain independence, and discover the satisfaction that comes from taking ownership of your car's maintenance.

So, let's get started. Together, we'll uncover the secrets that mechanics don't want you to know and turn car care into something you can tackle with confidence. Your journey to stress-free, cost-effective car ownership starts here.

CHAPTER 1

UNDERSTANDING HOW YOUR CAR WORKS

YOUR CAR ISN'T A MYSTERY—IT'S A MACHINE, AND YOU'RE ABOUT TO LEARN HOW IT TICKS.

A few years ago, my neighbor Sarah showed up at my garage, clearly frustrated. Her car had developed a strange rattling noise, and the mechanic had quoted her a hefty repair bill. She asked if I could take a quick look. It didn't take long to find the problem—a loose heat shield. A simple adjustment with a wrench solved it, saving her hundreds of dollars. She laughed in disbelief and said, "Why didn't anyone tell me cars aren't as scary as they seem?" That moment stuck with me, and it's the inspiration behind this chapter.

Cars can feel overwhelming if you don't understand how they work. But when you break them down into their core systems, they become surprisingly straightforward. The engine, brakes, and electrical components might seem like mysteries, but each plays a clear role in keeping you safe, comfortable, and on the move.

In this chapter, we'll unravel the basics of your car's key systems. You'll learn how the engine powers your vehicle, how the cooling system prevents overheating, and how the braking system ensures your safety. We'll also discuss the powertrain, electrical system, and suspension—explaining how these parts work together to deliver a smooth driving experience.

One of the most underrated tools in understanding your car is the owner's manual. That little book holds a wealth of information tailored to your specific vehicle. From maintenance schedules to troubleshooting tips, it's your go-to guide for keeping things running smoothly.

By the end of this chapter, you'll feel more confident about recognizing potential issues, speaking with mechanics, and even tackling simple fixes on your own. Let's take the mystery out of what's under the hood and empower you to take better care of your car.

THE BASICS OF VEHICLE SYSTEMS

Every car is like an orchestra, with each system playing a unique role in delivering performance, safety, and comfort. Instead of thinking of your vehicle as a single machine, imagine it as a team of experts working together to make every drive smooth and enjoyable. Understanding how these systems work can feel intimidating at first, but breaking them down makes them much more approachable. With a little knowledge, you'll not only gain confidence in diagnosing problems but also be better equipped to keep your car running in top shape.

Let's break it down, step by step, and dive into the core systems that bring your vehicle to life: the engine, powertrain, electrical system, cooling and heating systems, braking system, and suspension system.

Understanding the Engine: The Powerhouse of Your Car

If your car were a living organism, the engine would undoubtedly be its heart. For most vehicles, this heart takes the form of an internal combustion engine (ICE). It's a powerful piece of technology that transforms fuel into the energy needed to get your car moving. This process might sound complicated, but it's surprisingly simple when broken down:

1. **Intake:** The engine takes in air, mixing it with fuel in a precisely measured ratio.
2. **Compression:** A piston compresses this mixture inside the engine's cylinder, making it more potent.
3. **Combustion:** A spark plug ignites the compressed mixture, creating a controlled explosion that drives the piston downward.
4. **Exhaust:** The waste gases from this process are expelled through the exhaust system.

This cycle happens hundreds, sometimes thousands, of times every minute while you drive. That's a lot of hard work for the engine! To keep it running smoothly, regular oil changes are non-negotiable. Engine oil lubricates all those moving parts, reducing wear and tear and preventing overheating. It's also crucial to replace air and fuel filters as recommended in your car's manual to ensure the engine gets clean air and fuel. A well-maintained engine isn't just about performance—it's about longevity, saving you money on repairs down the road.

The Powertrain: Delivering Energy to the Wheels

If the engine generates power, the powertrain ensures it reaches the wheels, allowing your car to move. It's essentially the bridge between the engine and the road, and it includes several key components:

- **Transmission:** This adjusts the power coming from the engine to match your driving speed. Automatics do this seamlessly, while manual transmissions rely on you to shift gears.
- **Driveshaft:** Think of this as the delivery system that transports power from the transmission to the differential.
- **Differential:** This clever device ensures that power is distributed to the wheels correctly, allowing them to spin at different speeds when you turn.

The powertrain is vital for smooth and efficient movement. If something's off—like strange noises, difficulty shifting gears, or vibrations—it's time to take a closer look. Regularly checking your transmission fluid is one simple way to help keep this system in peak condition. A well-maintained powertrain not only enhances your driving experience but also ensures your car operates safely and reliably.

The Electrical System: Your Car's Nervous System

Your car's electrical system might seem invisible, but it plays a role in nearly every part of your vehicle's operation. From starting the engine to powering the headlights and charging your phone, it's the unsung hero of modern driving. Let's break it down into its core components:

1. **Battery:** This is your car's power reserve. It provides the energy needed to start the engine and keeps smaller systems running when the engine is off.
2. **Starter Motor:** Once the battery does its job, the starter motor kicks in, turning the engine over to get it running.
3. **Alternator:** As the engine runs, the alternator takes over, generating electricity to keep the battery charged and powering electrical systems.

But there's more to your car's electrical system than just these three components. Modern vehicles are equipped with an array of advanced features, such as:

- **Infotainment Systems:** From navigation to hands-free calling, these features keep you connected and entertained on the road.
- **Safety Sensors:** Tire pressure monitors, blind-spot detection, and collision warnings all rely on electrical systems.
- **Engine Control Unit (ECU):** This is the car's brain, using data from sensors to optimize engine performance and fuel efficiency.

To keep your electrical system in good health, make sure battery terminals are clean and corrosion-free, and check your alternator's output during routine maintenance. And always keep jumper cables or a portable jump starter handy—you never know when you might need them!

Cooling and Heating Systems: Managing Temperature

Your car's engine works hard, and all that effort generates a tremendous amount of heat. Without a cooling system, your engine would overheat in minutes. Here's how the cooling system keeps everything in check:

- **Coolant:** This liquid, a mix of water and antifreeze, absorbs heat from the engine as it circulates.
- **Radiator:** Once the coolant is heated, it flows into the radiator, where the heat is released into the air.
- **Thermostat:** This small but mighty part regulates the flow of coolant, ensuring the engine operates at the right temperature.

The heating system works alongside the cooling system, using some of the engine's heat to warm the cabin during colder months. If your heater isn't working, it might be a sign of low coolant levels or a thermostat issue.

Your car's cooling system keeps the engine from overheating, so regular care is essential. Check your coolant levels often, especially before long trips or in extreme weather. Low levels can lead to overheating and severe engine damage. Also, watch for leaks—green or orange puddles under your car signal trouble. Address leaks early to avoid costly repairs. Lastly, replace the coolant as recommended in your owner's manual. Old coolant can corrode components and cause clogs. Fresh coolant prevents rust and buildup, keeping your system efficient. A little attention goes a long way in protecting your engine and ensuring smooth performance.

Braking System: Your Car's Safety Net

The brakes are your first line of defense when it comes to safety. They convert the car's kinetic energy into heat, bringing it to a stop when needed. Here's how they work:

- **Brake Pedal:** Pressing the pedal activates the hydraulic system, increasing pressure.
- **Brake Pads and Rotors:** This pressure forces the brake pads against the rotors, creating friction that slows the car.

Modern cars often come equipped with Anti-lock Braking Systems (ABS), which prevent the wheels from locking up during sudden stops. This ensures you maintain control even in emergency situations.

If you hear squealing or grinding noises when braking, or if the pedal feels spongy, don't ignore it. These are signs your brakes need attention. Regularly check brake pads and rotors for wear and flush brake fluid as needed.

Suspension System: Ensuring a Smooth Ride

While the brakes help you stop, the suspension system ensures you stay comfortable and in control during your drive. It's responsible for absorbing shocks from the road and keeping your car stable. The main components include:

- **Springs:** These absorb energy from bumps and support the car's weight.
- **Shocks and Struts:** They work alongside springs to reduce vibrations and improve handling.
- **Control Arms and Joints:** These parts maintain wheel alignment for smooth motion.

A worn-out suspension system can make your car's ride rough and even unsafe. Pay attention to signs like excessive bouncing, pulling to one side, or uneven tire wear. Promptly addressing these issues can save you from more expensive repairs down the road.

Why Understanding These Systems Matters

Here's the thing: you don't need to become a professional mechanic to understand your car. A basic knowledge of these systems equips you to recognize when something isn't right, communicate effectively with mechanics, and even handle routine maintenance yourself. This knowledge empowers you to save money, avoid stress, and keep your car running reliably for years.

Remember, your car is more than just a machine—it's your partner in daily life, adventure, and everything in between. Taking care of it isn't just about performance; it's about building a relationship with a vehicle that gets you where you need to go safely and confidently. Treat it well, and it'll return the favor every time you hit the road.

THE OWNER'S MANUAL: UNLOCKING ITS FULL POTENTIAL

When was the last time you opened your car's owner's manual? If your answer is "never" or "not in a while," you're not alone. Most people overlook this little booklet, often leaving it to gather dust in the glove compartment. But here's the thing—it's one of the most useful tools you have for understanding and maintaining your car. Think of it as a personal guidebook specifically written for your vehicle to help you keep things running smoothly.

Why Your Owner's Manual Matters

Your owner's manual isn't just a random stack of pages thrown in with your car—it's a treasure trove of tailored information. Unlike generic advice you might come across online, this booklet is designed with your specific make and model in mind. Whether you're trying to understand the basics or troubleshoot a problem, it should always be your first stop.

For example, instead of guessing or relying on vague advice, the manual tells you exactly what type of oil your engine needs, how much air to put in your tires, and even when to schedule routine maintenance tasks like air filter changes and tire rotations. Having this kind of precise information at your fingertips can save you from costly mistakes and unnecessary visits to the mechanic.

What You Can Learn from Your Manual

The owner's manual covers a lot of ground, but here are a few sections you'll want to get familiar with right away:

- **Maintenance Schedules:** This is your car's version of a to-do list. It lays out when to tackle essential tasks like oil changes, brake inspections, and tire rotations. Following these recommendations keeps your car running smoothly and helps you avoid expensive repairs down the line.
- **Specifications:** Ever wondered what kind of oil to use or the exact tire pressure your car needs? The manual spells it out clearly, so you don't have to guess. It also provides details on replacement parts, like the correct size of windshield wipers or light bulbs.
- **Troubleshooting Guides:** Dealing with a minor issue like a warning light or jump-starting your car can feel overwhelming, but the manual simplifies it. These guides walk you through common problems step by step, making them much easier to handle.

The more you use your owner's manual, the more confident you'll become as a car owner. It takes the mystery out of car care and turns what might seem like complicated tasks into manageable ones. For example, when a dashboard light comes on, you won't panic—you'll know exactly where to look to figure out what it means and what to do next.

This knowledge also helps when you need to work with a mechanic. Instead of feeling lost or unsure, you'll be able to ask the right questions and understand the services being offered. That way, you can avoid unnecessary repairs or upsells and make informed decisions about your car's care.

To get the most out of your owner's manual, keep it somewhere handy, like the glove compartment, so it's always within reach. Whenever you have a question or notice something unusual with your car, take a moment to consult the manual before seeking help elsewhere. Over time, you'll find it becoming your go-to resource for all things car-related. And trust me, once you start relying on it, you'll wonder why you waited so long to open it.

RECAP

Cars are so much more than a way to get from point A to point B—they're intricate systems working together in perfect harmony. In this chapter, we explored those systems, from the engine and powertrain to the electrical and braking systems. Each one has a critical role in keeping your car safe, reliable, and efficient. Now, you know their basic functions, the signs that something might be wrong, and how regular maintenance is the key to keeping everything running smoothly.

We also highlighted one of the most overlooked yet valuable tools at your disposal: your car's owner's manual. It's not just a booklet gathering dust in your glove compartment—it's your car's personalized care guide. Whether it's choosing the right oil, decoding a mysterious dashboard light, or knowing when to schedule routine maintenance, the manual is packed with insights that can save you time, money, and frustration.

With this knowledge, you're now better equipped to take charge of your car's care. You don't need to be a professional mechanic to recognize warning signs, communicate effectively with repair shops, or make informed decisions. Just understanding the basics of how your car works puts you miles ahead, giving you confidence as a car owner.

Next, we'll move from knowledge to action by focusing on the tools you need to handle DIY car care. Whether you're a beginner putting together your first toolkit or an enthusiast ready to upgrade, we'll walk you through the essentials and a few advanced tools that make maintaining your car easier and more effective.

CHAPTER 2

TOOLS AND ESSENTIALS FOR DIY CAR CARE

THE RIGHT TOOLS MAKE EVERY CAR REPAIR FEEL LIKE SECOND NATURE

Did you know that nearly 90% of car maintenance tasks can be tackled with just a handful of basic tools? It's true. The difference between a costly trip to the mechanic and a quick DIY fix often comes down to whether you have the right tools on hand.

This chapter is all about building your toolkit, whether you're just starting out or looking to step up your game. We'll walk through the essentials—like screwdrivers, wrenches, and a reliable tire pressure gauge—that form the backbone of any DIY car care setup. You'll also learn about specialized tools, such as torque wrenches and OBD2 scanners, that can make even complex repairs feel manageable.

But having the tools is only half the story. We'll also discuss how to set up a functional workspace, maintain your equipment, and choose quality over quantity when shopping for tools. By the time you're done with this chapter, you'll know exactly what you need to take care of your car safely, efficiently, and with confidence.

Ready to turn your garage into your very own car care command center? Let's dive in.

BUILDING YOUR BASIC TOOLKIT FOR MAINTENANCE

Every DIY car enthusiast starts somewhere, and building your first toolkit is the foundation of successful car care. The right tools don't just make maintenance and repairs easier—they give you the confidence to tackle projects without relying on a mechanic. Think of your toolkit as an investment in your skills and independence. Let's walk through the essentials and how they'll help you get the job done.

Screwdrivers and Wrenches: The Core of Your Toolkit

Screwdrivers and wrenches are the bread and butter of any tool collection. A set of flathead and Phillips screwdrivers in various sizes will cover a wide range of tasks, from tightening hose clamps to replacing battery terminals or removing dashboard components. These are versatile, simple tools you'll find yourself reaching for again and again.

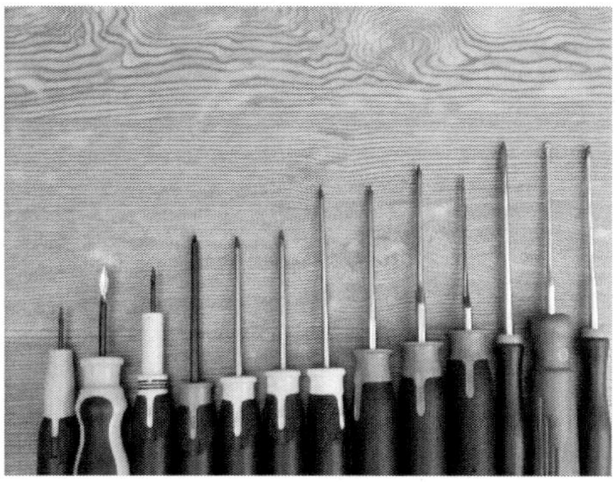

When it comes to wrenches, a combination set with both open-ended and box-end options is a must. They'll help you handle most nuts and bolts you encounter during car maintenance. Adding an adjustable wrench is also a great idea—it's perfect for tackling non-standard bolt sizes or plumbing connections, like those on coolant lines.

- **Pro Tip:** Go for quality when buying wrenches. Look for tools made from drop-forged steel with a chrome finish. These are durable, rust-resistant, and built to handle heavy use.

Socket Set: Your Go-To for Bolts

A good socket set is indispensable for car repairs. From engine components to suspension systems, most of your car is held together by bolts, and a socket wrench will help you remove and tighten them quickly and efficiently. Here's what to look for in a socket set:

- **Socket Sizes:** Get both metric and SAE (imperial) sizes since many vehicles use a mix of these standards.
- **Extensions:** These help you access hard-to-reach bolts, such as those in spark plug wells or under engine brackets.
- **Ratchet Handles:** A 3/8-inch drive is versatile for most tasks, while a 1/2-inch drive works well for heavier jobs.
- **Magnetic Sockets:** These prevent small bolts from falling into hard-to-reach areas, saving you time and frustration.

A socket set is one of those tools you'll use so often you'll wonder how you managed without it.

Pliers: Grip, Cut, and Control

Pliers are another must-have. Their versatility makes them essential for gripping, bending, cutting, and more. Needle-nose pliers are ideal for working with small wires or fuses, while slip-joint pliers handle general tasks like gripping hoses or tightening clamps. If you've ever struggled with a stuck bolt or clamp, locking pliers (often called Vise-Grips) will be your best friend—they offer a strong, steady grip when extra force is needed.

Planning to work on your car's electrical systems? Add wire cutters or combination pliers to your kit. These tools make it much easier to handle and cut wires when replacing fuses or repairing electrical connections.

Tire Pressure Gauge: Safety Meets Efficiency

Your tires are the only part of your car that touches the road, so maintaining the right tire pressure is essential for safety and efficiency. An accurate tire pressure gauge ensures your tires are inflated to the manufacturer's recommended levels. This not only extends the life of your tires but also improves fuel efficiency and reduces the risk of blowouts.

You can choose between pencil-style gauges (affordable and reliable) or digital ones, which are often more accurate and easier to read.

- **Pro Tip:** Check your tire pressure regularly, especially before long trips or during seasonal temperature changes. Cold weather can lower tire pressure, making this a critical part of winter car care.

Other Essentials to Keep on Hand

While screwdrivers, wrenches, and pliers form the backbone of your toolkit, a few additional items can make your DIY car care experience smoother and safer:

- **Gloves:** Protect your hands from cuts, burns, and exposure to grease and chemicals. Look for gloves with a good grip and oil resistance.
- **Flashlight or Work Light:** A rechargeable LED work light is invaluable when working in dim areas, like under the hood or beneath your car. Hands-free options, such as headlamps or magnetic lights, are especially handy.
- **Utility Knife:** This tool is perfect for cutting hoses, belts, or electrical tape. A sharp, retractable blade ensures safety and precision.
- **Rags or Shop Towels:** Keep these on hand for wiping up oil, grease, or coolant. Microfiber towels are an excellent reusable option.

It's easy to feel overwhelmed when shopping for tools, given the endless brands and options out there. Start simple: prioritize quality over quantity. Inexpensive tools might save you money upfront, but they often break or wear out quickly, costing more in the long run. Stick with trusted brands like Craftsman, Stanley, or DeWalt, which offer durable, reliable tools at reasonable prices.

Consider buying tools individually instead of opting for large all-in-one sets. This way, you can focus on acquiring high-quality essentials and avoid paying for tools you may never use. Don't forget to check for sales or shop secondhand—garage sales and online marketplaces can be treasure troves for budget-friendly, lightly used tools.

MUST-HAVE TOOLS FOR DIY REPAIRS

Once you've mastered the basics of car maintenance, it's only natural to want to take on bigger challenges. Projects like changing your oil, replacing brake pads, or inspecting your car's undercarriage can be incredibly satisfying (not to mention cost-saving), but they do require the right tools. The good news? With just a few specialized items, you can handle these tasks safely and efficiently. Let's break down the essentials.

Jack and Jack Stands: The Foundation of Safe Repairs

If you're planning to work underneath your car—whether to change a tire, check the undercarriage, or repair the brakes—you'll need two essential tools: a hydraulic floor jack and jack stands. The floor jack is responsible for lifting the car, while the jack stands ensure it stays securely supported while you work. This duo is absolutely essential for your safety and peace of mind.

When shopping for a floor jack, focus on a smooth lifting mechanism and make sure it has enough weight capacity for your vehicle. For cars that sit lower to the ground, like sports cars or some sedans, a low-profile jack is a great choice since it can slide under the car without any hassle. Jack stands, on the other hand, are all about stability. Look for stands with a wide base and a weight rating that exceeds your vehicle's weight to ensure they can handle the load securely.

- **Safety Tip:** Never, under any circumstances, rely on the jack alone to hold up your car. Always use jack stands to support the weight, and double-check their positioning before you get under the car. Those extra seconds spent securing your setup can make all the difference in keeping you safe.

Oil Filter Wrench: Simplifying Oil Changes

Changing your own oil is one of the easiest ways to start learning DIY car care. It's straightforward, cost-effective, and gives you a real sense of accomplishment. However, if you've ever struggled to loosen a stubborn oil filter, you know it can turn into a frustrating mess. That's where an oil filter wrench comes in—it's a simple tool that makes the process much smoother.

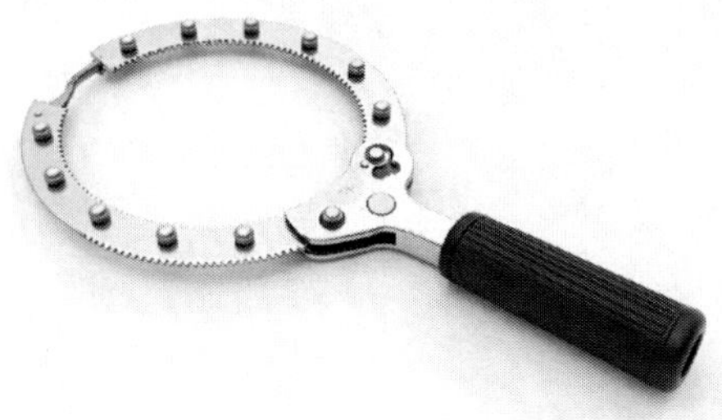

The oil filter wrench is designed to grip the filter securely, allowing you to remove it easily, even if it was tightened a bit too much last time. Without the right tool, you risk damaging the filter housing or spilling oil everywhere, which can turn an easy job into a big hassle. On the flip side, with the right oil filter wrench, oil changes become a quick, hassle-free part of your car care routine.

So, how does one choose the right wrench? If you're maintaining just one car, go for a wrench specifically sized for your vehicle's oil filter. On the other hand, if you'll be working on multiple vehicles or helping out friends and family, an adjustable oil filter wrench is worth the investment. Its versatility saves you time and money, making it a handy addition to your toolkit.

Funnel and Drain Pan: Mess-Free Fluid Changes

Fluid changes—whether it's oil, coolant, or transmission fluid—can be messy, but they don't have to be. With the right tools, you can keep the process clean, manageable, and stress-free. Two simple but essential tools that make all the difference are a funnel and a drain pan. These unsung heroes of car care can turn what might seem like a daunting task into a smooth, spill-free experience.

- **The Funnel:** Let's start with the funnel. This handy tool is perfect for neatly pouring fluids into tight spaces like the oil fill cap or radiator opening. Without one, you're almost guaranteed to spill, leaving you with an unnecessary cleanup. For better control and precision, choose a funnel with a long, narrow neck that can easily reach those tricky spots. It's a small investment that saves a lot of hassle.

- **The Drain Pan:** Now onto the drain pan. As your car's old fluids drain out, this tool catches them efficiently, keeping your workspace tidy. To make fluid disposal easier, look for a drain pan with a built-in spout. This feature lets you pour used fluids into recycling containers without any additional mess.

Here's a pro tip: opt for a drain pan with at least a 10-quart capacity. It's better to have extra room than risk an overflow, especially during larger fluid changes. With both a funnel and a drain pan in your toolkit, you'll be ready to handle fluid changes like a pro—without the mess.

Multimeter: Troubleshooting Electrical Issues

Electrical problems in cars can be intimidating, but they don't have to be. A multimeter is a simple tool that makes diagnosing these issues much easier. With it, you can check your car battery's voltage, measure the alternator's performance, and even trace problems in the wiring system. If you've ever felt stuck guessing what's wrong, a multimeter is your go-to for quick and accurate answers.

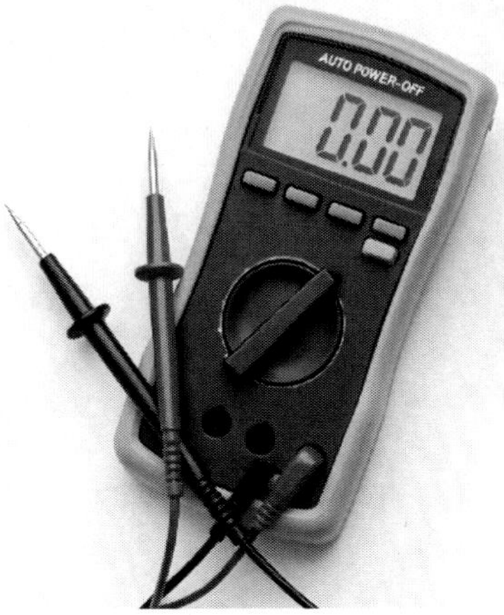

Electrical issues are surprisingly common, and without the right tools, troubleshooting can feel like shooting in the dark. A multimeter eliminates that guesswork. For example, if your car refuses to start, a few quick tests can reveal whether the culprit is a drained battery or something more complex, like a faulty alternator. Having this tool on hand means you're equipped to diagnose problems on the spot, saving you time, money, and frustration.

If you're new to electrical work, a digital multimeter is a great choice. It's beginner-friendly, with a clear display and automatic range selection to make readings simple. Look for one with the essential functions—voltage, resistance, and current testing. These features will cover most of your car's electrical troubleshooting needs.

ADVANCED TOOLS FOR MORE COMPLEX JOBS

If you've mastered the basics and feel ready to tackle more advanced repairs, it's time to consider adding a few specialized tools to your collection. These tools can help you take on jobs you might otherwise leave to a mechanic, saving you time and money and giving you the confidence that comes with doing it yourself.

Torque Wrench: Precision Tightening Made Easy

When working with bolts that need to be tightened to a specific level, a torque wrench is invaluable. Over-tightening can strip threads, while under-tightening might leave components dangerously loose. With a torque wrench, you can ensure bolts are tightened to the manufacturer's exact specifications.

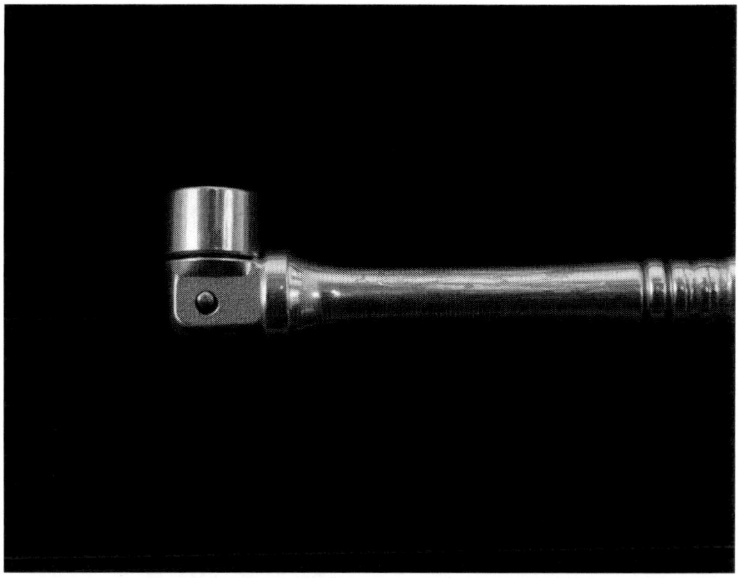

This tool is especially important for tasks like installing spark plugs, securing wheel lug nuts, or tightening suspension components. For example, properly torqued lug nuts protect your wheel hub from damage and keep your wheels securely in place.

When shopping, look for a torque wrench with a wide range of settings and an easy-to-read scale. Digital models provide precise readings but are often pricier, while mechanical wrenches are budget-friendly and equally reliable.

Code Reader or OBD2 Scanner: Diagnosing Like a Pro

Modern cars are equipped with onboard diagnostic (OBD) systems that generate error codes whenever something goes wrong. An OBD2 scanner plugs into your car's diagnostic port and reads these codes, making troubleshooting far simpler.

For instance, if your check engine light comes on, a quick scan can tell you if the issue is as minor as a loose gas cap or something more serious. Advanced OBD2 scanners offer live data like engine RPM, coolant temperature, and fuel efficiency, allowing you to monitor your car's health in real time.

Basic scanners are affordable and perfect for identifying error codes. If you're looking for more in-depth functionality, consider a model with live monitoring or Bluetooth connectivity that works with a smartphone app.

Impact Wrench: Tackle Stubborn Bolts with Ease

We've all encountered bolts that won't budge no matter how much effort we put in. That's where an impact wrench shines. This high-torque tool makes removing or tightening stubborn bolts, like those on wheels or suspension parts, a breeze.

For home use, electric impact wrenches are portable and easy to handle. Pneumatic models, which deliver even more torque, are great for advanced setups but require an air compressor.

- **Pro Tip:** When choosing an electric impact wrench, focus on torque output and battery life for cordless models. If you go pneumatic, ensure your air compressor can meet the tool's requirements for optimal performance.

Brake Bleeder Kit: Essential for Brake Maintenance

If you're planning to work on your car's braking system, a brake bleeder kit is a must-have. This tool removes air bubbles from the brake lines, ensuring your brakes function properly. Air in the lines can compromise braking performance, making it unsafe to drive.

Brake bleeder kits make flushing old fluid and refilling the system much easier. Look for kits with multiple adapters to fit different brake systems and a vacuum pump for efficient bleeding. Some kits even include clear instructions, making the process straightforward for beginners.

Compression Tester: Digging into Engine Diagnostics

A compression tester is your go-to tool for diagnosing internal engine issues. It measures the pressure in your engine's cylinders, helping you identify potential problems like worn piston rings, damaged valves, or a blown head gasket.

If your car is experiencing performance issues such as reduced power or rough idling, a compression test can reveal whether the problem lies within the engine. It's an essential tool for anyone diving deeper into engine diagnostics.

When selecting a compression tester, ensure it's compatible with your engine type and includes multiple adapters for different spark plug sizes.

SAFETY GEAR AND BEST PRACTICES

Working on your car can be incredibly rewarding—not to mention a great way to save money—but safety should always come first. With the right preparation, gear, and habits, you can make sure every repair or maintenance task is as smooth and safe as possible. Let's dive into the key steps to set yourself up for success.

Personal Protective Equipment (PPE)

Think of safety gear as an essential part of your toolkit. It's not just about avoiding injuries—it's about working comfortably and confidently. Here are a few must-haves for any DIY car project:

- **Gloves:** Nitrile gloves protect your hands from chemicals like oil and brake fluid, while heavy-duty gloves are great for mechanical tasks involving sharp or rough components.
- **Safety Glasses:** Protect your eyes from flying debris, splashing fluids, or unexpected sparks. They're a simple but vital piece of equipment.
- **Ear Protection:** Tools like impact wrenches and air compressors can be loud enough to damage your hearing over time. Quality earplugs or earmuffs are an easy fix.
- **Respirator Mask:** If your project involves fumes, dust, or particles—such as during a brake job or spray painting—wearing a respirator mask is a must.

Investing in PPE is a small price to pay for staying safe and focused on the job.

Setting Up Your Workspace

Even if you don't have a dedicated garage, you can still create an efficient and organized workspace for car maintenance. The key is to make the most of the space you have.

Start by designating a specific area—a corner of your garage, a driveway, or even a portable workbench. Use pegboards or magnetic strips to hang frequently used tools within easy reach. For smaller items like screws, bolts, and fuses, labeled containers or clear storage boxes can keep things tidy and easy to find.

Lighting is another crucial element. Good visibility can make the difference between a smooth repair and a frustrating mistake. A combination of overhead lights and portable LED work lights ensures you can see clearly, even in dim or cramped spaces. Finally, keep your workspace clean and clutter-free. This not only makes the job easier but also reduces the risk of accidents.

Maintaining Your Tools

Your tools are the backbone of your DIY efforts, and taking care of them is just as important as using them. After every use, clean off grease, grime, or dirt to keep them in top shape. Store them in a dry, secure location to prevent rust, and periodically inspect them for wear or damage.

Consider investing in a sturdy toolbox or tool chest to keep everything organized. Not only will this save you time when looking for the right tool, but it'll also protect your equipment, ensuring it lasts for years. Well-maintained tools aren't just more reliable—they're also safer to use.

One more thing: when it comes to building your toolkit, quality beats quantity every time. While cheaper tools might seem like a bargain upfront, they often wear out quickly or fail when you need them most. Instead, invest in trusted brands like Craftsman, DeWalt, or Kobalt. These companies offer durable tools that perform well without breaking the bank.

- **Pro Tip:** Start small. You don't need to buy an entire toolset all at once. Focus on acquiring individual high-quality tools as you need them. You can also find great deals on secondhand tools at garage sales or online marketplaces—just inspect them carefully for damage before buying.

Preparation is half the battle when it comes to DIY car care. Here are a few more tips to help you get started on the right foot:

- **Invest in a Service Manual:** A vehicle-specific service manual provides detailed instructions and diagrams tailored to your car. It's like having a professional mechanic guide you step-by-step.
- **Start Small:** If you're new to car care, begin with simple tasks like checking tire pressure, replacing wiper blades, or changing air filters. As your confidence grows, move on to more complex repairs.
- **Stock Common Parts:** Keep a supply of commonly replaced items—fuses, light bulbs, and filters. Having these on hand saves time and prevents unnecessary trips to the store.
- **Learn from Others:** Join online forums or watch instructional videos to expand your knowledge. Many experienced DIYers share tips and tricks that can make your projects easier and more efficient.

DIY car care is as much about preparation and safety as it is about skill. By investing in quality tools, setting up an organized workspace, and prioritizing safety gear, you can approach every task with confidence. Not only will you save money, but you'll also enjoy the satisfaction of taking care of your car yourself—and keeping it in great shape for years to come.

RECAP

With the right tools, you're already well on your way to mastering DIY car care. In this chapter, we covered the essentials that every car owner needs. From screwdrivers and wrenches to more advanced gadgets like torque wrenches and OBD2 scanners, these tools don't just make repairs easier—they give you the confidence to tackle jobs you might have once thought were out of reach.

We also highlighted why an organized workspace matters. Whether it's a corner of your garage or a spot in your driveway, having a dedicated area can make all the difference. Keeping your tools clean and properly stored isn't just about tidiness; it ensures they're ready to go when you need them most. Plus, a little maintenance on your tools means they'll last for years, saving you money in the long run.

Starting out can feel overwhelming, but we stressed the importance of taking it slow. Focus on building your toolkit piece by piece, investing in quality items that will serve you reliably over time. Whether you're shopping new or secondhand, there's no rush—your collection will grow alongside your skills.

Now that you've got your tools in order, it's time to put them to good use. In the next chapter, we'll focus on the habits and maintenance routines that keep your car running smoothly and help you avoid those costly surprises. Let's get to it!

CHAPTER 3

PREVENTATIVE MAINTENANCE MASTERY

A LITTLE PREVENTATIVE CARE TODAY SAVES YOU FROM COSTLY REPAIRS TOMORROW

Have you ever thought about what really happens when you skip regular car maintenance? Sure, your car might seem fine for a while—it's still running, right? But even minor oversights, like putting off an oil change or brushing off a warning light, can lead to bigger problems. And those problems often show up at the worst possible time, like when you're on a road trip or late for work.

This chapter is here to help you avoid those frustrating, often expensive surprises. We'll break down the basics of preventative maintenance into simple, manageable steps. You'll learn how routine tasks like checking your fluids, inspecting your tires, and replacing filters can add years to your car's life. We'll also talk about battery care so you're not stuck with a car that won't start when you need it most.

The good news? These tasks aren't complicated, and they don't take much time. A little effort now can save you a lot of money and stress later. You'll find that regular maintenance not only keeps your car reliable but also boosts fuel efficiency and ensures your safety on the road.

By the end of this chapter, you'll feel confident in your ability to take care of your car. These small habits might seem like no big deal, but they add up to a vehicle that's always ready to go when you are. Let's dig into the routines that will keep your car running smoothly, one mile at a time.

THE IMPORTANCE OF REGULAR MAINTENANCE

Taking care of your car goes beyond keeping it running smoothly—it's about protecting your investment, staying safe, and saving money over time. Think of preventative maintenance as your car's version of a regular check-up with the doctor. Catching small issues early is always easier (and cheaper) than dealing with major problems later.

Skipping maintenance might seem harmless at first, but it can lead to costly repairs down the line. For instance, ignoring a low oil level could result in severe engine damage, while neglecting tire rotations might cause uneven wear, reducing fuel efficiency and affecting handling. These simple tasks might feel minor, but they make a big difference in how well your car performs and how long it lasts.

Reliability is another key benefit of regular maintenance. A car that's well cared for is far less likely to break down when you least expect it. No one wants to be stranded on the side of the road because of a preventable issue. Routine care minimizes these risks and keeps your vehicle ready to go whenever you need it.

Beyond safety and reliability, maintenance also adds value to your car. If you ever plan to sell or trade it in, having a record of regular care can make your car far more appealing to buyers. It shows that you've taken good care of it, which can boost its resale price and give potential buyers peace of mind.

The best part? Preventative maintenance doesn't have to be overwhelming. Focus on the essentials, like oil changes, tire care, battery checks, and fluid levels. These routine tasks help you spot problems early, saving you from expensive surprises down the road. In the end, maintenance isn't just a task—it's an investment in your car's longevity and your own peace of mind.

CHECKING AND CHANGING FLUIDS

Fluids are the lifeblood of your car, keeping all its essential systems running smoothly. Whether it's lubricating the engine, regulating temperatures, or ensuring proper braking, each fluid plays a critical role. Regularly checking and changing them is one of the easiest ways to maintain your car's performance and avoid costly repairs. Let's walk through the process, step by step.

Engine Oil: The Lifeline of Your Engine

Engine oil is arguably the most important fluid in your car. It reduces friction, lubricates moving parts, and prevents overheating. However, over time, oil collects dirt and debris, which can cause wear and tear if not replaced regularly.

- **How to Check Engine Oil:** Start by parking your car on a level surface and letting the engine cool. Open the hood and locate the dipstick, usually marked with a bright handle. Pull it out, wipe it clean, and reinsert it fully. Then, pull it out again to check the oil level and color. Clean oil should be amber and transparent, while dirty oil appears dark and gritty.

- **When to Change Engine Oil:** Refer to your owner's manual for the recommended interval, usually between 3,000 and 7,500 miles. If you frequently take short trips or tow heavy loads, consider changing it more often.

Lastly, ensure you use a funnel when adding oil to avoid spills, and always recycle old oil at an authorized facility.

Coolant: Regulating Engine Temperature

Coolant, or antifreeze, prevents your engine from overheating in summer and freezing in winter. Driving with low or dirty coolant can lead to severe engine damage, so keeping it in check is essential.

- **How to Check Coolant:** Locate the coolant reservoir—it's a translucent tank with "min" and "max" level indicators. Ensure the fluid is between these lines. If it's low, add the coolant specified in your owner's manual. Always wait until the engine is cool before opening the radiator cap to avoid burns.
- **When to Flush Coolant:** Coolant loses effectiveness and accumulates debris over time. Flush and replace it every 30,000 to 50,000 miles or as recommended by your manual.

Transmission Fluid: Ensuring Smooth Shifting

Transmission fluid helps your gears shift smoothly and protects the transmission from overheating. While it doesn't need frequent changes, regular checks are important.

- **How to Check Transmission Fluid:** Some cars have a dipstick for transmission fluid, usually near the back of the engine bay. With the engine running and the car in park, pull out the dipstick, wipe it clean, reinsert it, and pull it out again. Check the fluid level and color—clean transmission fluid is pinkish-red, while dirty fluid looks dark or burnt.
- **When to Change Transmission Fluid:** Replace it every 30,000 to 60,000 miles or as specified in your owner's manual.

Other Fluids to Monitor

While engine oil, coolant, and transmission fluid are the big players, several other fluids deserve regular attention:

- **Brake Fluid:** Brake fluid ensures your brakes respond promptly. Over time, it absorbs moisture, which reduces its effectiveness. Inspect the reservoir every six months and replace the fluid every two to three years or if it appears dark.

- **Power Steering Fluid:** This fluid keeps your steering smooth and responsive. Check it monthly by inspecting the reservoir, usually labeled near the engine bay. Top it off if it's low, and watch for signs of leaks.
- **Windshield Washer Fluid:** Though often overlooked, windshield washer fluid is vital for clear visibility. Check it regularly and top it off as needed, especially during rainy or dusty conditions.

Staying on top of fluid maintenance might seem like a small task, but it's one of the most effective ways to avoid major issues. Low or dirty fluids can lead to overheating, brake failure, or costly mechanical problems. By proactively checking and changing fluids, you're not just protecting your car—you're saving yourself time, money, and headaches.

FILTERS: KEEPING YOUR CAR BREATHING CLEANLY

Filters are like your car's lungs—they keep harmful contaminants out and ensure clean air flows where it's needed. Whether it's protecting your engine or improving the air quality inside your car, filters play a vital role in keeping your vehicle running smoothly. Over time, though, these filters can become clogged, reducing performance and comfort. Fortunately, inspecting and replacing them is easy and makes a big difference.

Engine Air Filter: Protecting Your Engine

Your engine air filter keeps dirt, dust, and debris from entering the engine. A clean filter not only improves performance and fuel economy but also helps extend the life of your engine. When the filter becomes clogged, it restricts airflow, forcing your engine to work harder and burn more fuel—an unnecessary strain on both your car and your wallet.

- **When to Replace:** It's a good idea to inspect your engine air filter every 15,000 to 30,000 miles. If you drive on dusty roads, you may need to check it more often. A quick glance will usually tell you if it's time for a replacement—if the filter looks dirty or clogged, it's time to swap it out.
- **How to Replace:**
    1. Locate the filter housing, usually near the engine, often marked for easy identification.
    2. Open the housing and remove the old filter.
    3. Compare the old filter with the new one to ensure it's the correct size and type.
    4. Insert the new filter, close the housing, and secure it.

This is a quick, tool-free task that takes just minutes but has a noticeable impact on your engine's performance.

Cabin Air Filter: Clean Air for You and Your Passengers

Your cabin air filter ensures that the air inside your car is clean by trapping dust, pollen, and pollutants. This is especially important if you or your passengers have allergies. A clogged cabin filter can reduce airflow, strain your HVAC system, and even cause unpleasant odors.

- **When to Replace:** Plan to replace your cabin air filter every 12,000 to 15,000 miles or at least once a year. If you notice reduced airflow, musty smells, or more dust in the cabin, it's likely time for a new filter.
- **How to Replace:** Most cabin air filters are located behind the glove compartment. Here's how to replace it:
  1. Empty the glove compartment and gently detach its hinges to lower it.
  2. Locate the filter cover, remove it, and slide out the old filter.
  3. Insert the new filter, making sure the airflow arrows are aligned correctly.
  4. Reattach the cover and glove compartment.

Neglecting your car's filters can lead to reduced performance, lower fuel efficiency, and added strain on critical systems like your engine and HVAC. Regularly replacing both the engine and cabin air filters is a simple, affordable way to protect your car and ensure a comfortable ride for everyone.

By keeping your filters clean and up to date, you're giving your car the ability to "breathe" freely, which helps both the engine and passengers stay healthy and comfortable. Add these tasks to your regular maintenance routine, and you'll save time and money and avoid unnecessary stress down the road.

TIRE CARE: PRESSURE, ROTATION, AND TREAD INSPECTION

Your tires are your car's only contact with the road, making their condition critical for both safety and performance. Well-maintained tires provide a smoother ride, better fuel efficiency, and reduce the risk of blowouts or accidents. Let's dive into the three key aspects of tire care: pressure, rotation, and tread inspection, and why they're essential for keeping your car road-ready.

Checking Tire Pressure: A Simple Yet Crucial Task

Keeping your tire pressure in check isn't just about having a smooth ride—it plays a big role in how long your tires last, how much fuel your car uses, and how well it handles on the road. Ignoring tire pressure can cost you in more ways than one, like burning through fuel faster, wearing out your tires unevenly, or even risking a dangerous blowout. When your tires are underinflated, they create extra friction, which speeds up wear and tear. On the other hand, overinflated tires can reduce your grip on the road and make every bump feel sharper.

Here's how you can check your tire pressure:

1. Start with cold tires. Tire pressure increases as you drive, so for the most accurate reading, check the pressure when the tires are cold.
2. Locate the recommended PSI (pounds per square inch) for your car. You'll find this in your owner's manual or on a sticker inside the driver's side door frame.
3. Use a tire pressure gauge to measure the PSI for each tire.
4. If the pressure is too low, add air using an air compressor until it matches the recommended level. If it's too high, release air by gently pressing the valve.

Make it a habit to check your tire pressure at least once a month and always before long road trips. A digital tire pressure gauge is affordable, easy to use, and provides quick, accurate readings. Many gas stations also have air pumps with built-in gauges, but it's helpful to have your own for convenience.

Tire Rotation: Spreading the Wear Evenly

Your front and rear tires wear differently due to factors like braking, acceleration, and steering. Over time, this uneven wear can lead to reduced performance, shortened tire life, and even safety concerns. That's why regularly rotating your tires is so important. By distributing wear evenly, you can extend their lifespan, maintain balanced handling, and even improve fuel efficiency.

While rotating your tires, take a moment to check for uneven wear patterns. These can be early warning signs of alignment issues or suspension problems that might need professional attention.

It's also a good idea to check your tire tread depth while you're at it. A simple DIY trick is to use a coin: insert it into the tread grooves with the top of the coin facing downward. If you can see the top of the figure (like Lincoln's head on a penny or Washington's head on a quarter), it's time to replace your tires.

It's recommended to rotate your tires every 5,000 to 7,500 miles or during each oil change. Your owner's manual might provide specific instructions or rotation patterns, particularly for all-wheel-drive or performance vehicles. If you're ready to take on this task yourself, here's how to do it step by step:

1. **Gather your tools:** You'll need a jack, jack stands, a lug wrench, and a torque wrench if possible.
2. **Loosen the lug nuts:** Before lifting the car, slightly loosen the lug nuts on all four tires.
3. **Lift the car:** Use the jack to lift the car and secure it with jack stands. Never work under a car supported only by a jack.
4. **Swap the tires:** Follow your vehicle's recommended rotation pattern. Common patterns include front-to-back or cross-switching the rear tires to the front.

5. **Tighten the lug nuts:** Hand-tighten them first, then lower the car and use a torque wrench to secure them to the manufacturer's specifications.

If you're not comfortable rotating your tires at home, a professional service can handle it for you. They'll also ensure proper alignment and torque, which are essential for optimal tire performance.

Inspecting Tread Depth: Keeping Your Grip on the Road

Your tires' tread is what keeps you safely connected to the road, especially in tricky conditions like rain, snow, or ice. When the tread wears down, your tires lose their ability to grip the road, increasing the chances of skidding or losing control. That's why it's so important to keep an eye on your tread depth—it's a small habit that can make a big difference in your safety and your car's performance.

The good news? Checking your tread depth is quick, easy, and doesn't require any special tools. The penny test is a tried-and-true method:

1. Insert a penny into the tire tread with Lincoln's head facing down.
2. If you can see the top of Lincoln's head, the tread is too worn, and it's time to replace the tire.

It's a good idea to check your tread depth once a month and after long trips. While most states require a minimum tread depth of 2/32 of an inch, it's safer to replace your tires before they get that worn—especially if you drive in wet or icy conditions where extra traction is crucial.

But tread depth isn't the only thing to monitor. Uneven wear patterns can signal underlying issues with your car. For instance, excessive wear on one side of a tire might mean your alignment is off, while scalloped or cupped tread could point to suspension problems. Catching these signs early not only helps you avoid bigger issues but also keeps your car running smoothly and safely.

Regular tread inspections take just a few minutes but can save you a lot of trouble down the road. By staying proactive, you'll ensure your tires are always ready to handle whatever the road throws at you.

Before we move on, it's important to note that tires are an investment, and caring for them doesn't just save you money—it keeps you safer on the road. By keeping your tires properly inflated, rotated, and inspected, you can extend their lifespan, improve your car's performance, and reduce the likelihood of accidents or breakdowns. Plus, a little preventative care now means fewer headaches later.

Set aside a few minutes each month to check your tire pressure, rotate your tires at the recommended intervals, and inspect the tread. These small steps go a long way toward keeping your car running smoothly and ensuring peace of mind on every journey.

BATTERY MAINTENANCE FOR LONG-TERM RELIABILITY

Your car's battery is the heart of its electrical system. Without it, the engine won't start, leaving you stranded at the worst possible moment. Fortunately, with regular maintenance, you can ensure your battery remains reliable and avoid the stress of sudden failures.

Regular Inspection: Catch Issues Early

A quick glance at your battery can help you catch problems before they escalate.

Start with the terminals. Corrosion often builds up around them, appearing as a white or bluish powder. If left unchecked, this buildup can interfere with the electrical connection. Cleaning it is simple—just mix baking soda with water to create a paste, apply it with a small brush, and rinse it off with clean water. Once the terminals are dry, you're good to reconnect.

Next, check the connections. Loose or poorly connected terminals can cause intermittent electrical issues. Make sure the cables are secure, but avoid overtightening, as this can damage the battery posts.

Testing and Replacement: Know When It's Time

Testing your battery regularly helps you stay ahead of potential problems. Use a multimeter to check the voltage—12.6 volts when the car is off indicates a healthy battery. If the reading is consistently lower, it might be time to start considering a replacement.

Most car batteries last three to five years, though climate and driving habits can affect their lifespan. Watch for warning signs, like slow engine starts, dimming headlights, or struggling power windows. If you notice these symptoms, don't wait for the battery to fail completely. Replacing it proactively can save you from an inconvenient breakdown.

Battery Care Tips

A few simple habits can help extend your battery's life and keep it running efficiently:

- **Avoid Draining the Battery:** Turn off lights, chargers, and electronics when the engine is off to prevent unnecessary drain.
- **Use a Trickle Charger:** If your car sits idle for long periods, a trickle charger can maintain the battery's charge, keeping it in good condition.
- **Mind the Weather:** Extreme temperatures can stress your battery. In cold climates, a battery blanket can keep it from freezing, while in hot weather, parking in the shade can help reduce heat-related wear.

A healthy battery isn't just about convenience—it's about peace of mind. Regular checks and maintenance help you avoid surprises, extend the battery's life, and ensure your car is always ready to go when you need it. By spending just a few minutes each month on battery care, you're making a small investment in long-term reliability.

Think of it as one less thing to worry about in your day-to-day life. Preventative maintenance saves you time, money, and stress, giving you confidence that your car will start reliably—no matter what the day has in store.

RECAP

Preventative maintenance is the ultimate way to keep your car reliable, efficient, and safe. In this chapter, we focused on the simple but essential tasks that make a big difference—like checking your fluids, rotating tires, and replacing filters. These little efforts don't take much time, but they go a long way in avoiding expensive surprises and keeping your car running smoothly.

We also talked about the importance of keeping an eye on your battery, inspecting your tires regularly, and monitoring key systems like your brakes and cooling. These aren't just chores—they're smart investments that protect your car and give you peace of mind every time you get behind the wheel.

With these routines under your belt, you're already ahead of the game. You've taken the first steps toward being proactive about your car's care, which means fewer breakdowns, more savings, and a smoother ride.

Now, it's time to level up. In the next chapter, we'll dive into the art of understanding what your car is trying to tell you. Whether it's a strange noise, a dashboard light, or something that just feels "off," we'll teach you how to troubleshoot like a pro and tackle issues before they turn into bigger problems.

CHAPTER 4

DIAGNOSING PROBLEMS LIKE A PRO

YOUR CAR SPEAKS THROUGH WARNING LIGHTS—LEARN TO DECODE THEM LIKE A PRO

A friend of mine once called in a panic because her car wouldn't start, and a tow truck was on its way. I asked her a few questions: "Any strange noises? Lights on the dashboard?" She mentioned a "check engine" light had been glowing for weeks, but she thought it was nothing serious. When the mechanic got to it, they found a small problem with the spark plugs—something she could have fixed herself for under $20. That $20 fix turned into a $200 repair bill.

This chapter is about avoiding situations like that. Your car is constantly communicating with you through dashboard warning lights, odd noises, strange smells, and even vibrations. Understanding these signals can save you time, money, and frustration.

You'll learn to interpret dashboard warning lights like a pro, figure out what's causing those mysterious noises, and identify common problems through simple visual inspections. We'll also dive into using an OBD-II scanner, a handy tool that unlocks your car's error codes and tells you exactly what's going on under the hood.

By the end of this chapter, you'll have the confidence to decode your car's behavior, spot issues early, and decide whether it's a quick DIY fix or time to call a mechanic. Let's make sure you're never caught off guard by your car again.

WHAT YOUR CAR IS TELLING YOU: DASHBOARD WARNING LIGHTS

Dashboard warning lights are your car's way of speaking to you. Think of them as its built-in SOS signals, alerting you to potential issues before they snowball into major (and expensive) problems. While they might seem intimidating at first, understanding what these lights mean can save you time, stress, and money. Let's break them down one by one.

Check Engine Light: The Most Feared Warning

The check engine light (CEL) is the one most drivers dread seeing, but it doesn't always signal disaster. It can indicate a range of issues, from something as simple as a loose gas cap to more complex problems like a failing catalytic converter or oxygen sensor.

- **What to Do:** If the light is steady, you can safely drive short distances while investigating the issue. Start by checking the gas cap—it might just need tightening. However, if the light is flashing, stop driving immediately. A flashing CEL usually signals a severe problem, such as an engine misfire, which can cause significant damage if ignored.
- **Pro Tip:** Use an OBD-II scanner to read the error code associated with the CEL. These scanners are easy to use and invaluable for pinpointing the issue.

Oil Pressure Warning: A Signal to Stop Driving

The oil pressure warning light, shaped like a tiny oil can dripping liquid, is one you should never ignore. Low oil pressure means your engine isn't getting the lubrication it needs, which can lead to catastrophic damage.

- **What to Do:** Pull over as soon as it's safe and check your oil level. If it's low, topping it off might resolve the issue. However, if the light remains on or the problem recurs, it could point to a more serious issue, like an oil leak or failing oil pump. In such cases, consult a professional mechanic to avoid further damage.

Battery Warning: A Potential Power Problem

The battery warning light, shaped like a car battery, alerts you to issues with your car's electrical charging system. This could be due to a failing alternator, loose or corroded battery terminals, or even a dying battery.

- **What to Do:** If this light comes on while driving, your car will likely keep running until the battery's charge is depleted. To avoid getting stranded, check the battery and alternator as soon as possible. Start by inspecting the battery terminals for corrosion and ensuring the cables are tightly secured.

Coolant Temperature Warning: Prevent Overheating

The coolant temperature warning light, often depicted as a thermometer in liquid, signals that your engine is overheating—a problem that can cause serious damage like a blown head gasket or warped engine components.

- **What to Do:** Pull over immediately and allow the engine to cool before opening the hood. Check the coolant level in the reservoir—low coolant might be the issue. If you spot leaks or suspect a problem with the water pump or thermostat, get the car inspected promptly.

Tire Pressure Monitoring System (TPMS): Keep Tires Safe

The TPMS light, shaped like a horseshoe with an exclamation point, warns of low tire pressure. Driving on underinflated tires can reduce fuel efficiency, cause uneven wear, and increase the risk of a blowout.

- **What to Do:** Use a tire pressure gauge to check the pressure in all four tires. Inflate them to the manufacturer's recommended PSI, which you'll find on a sticker inside the driver's door or in the owner's manual. If the light remains on after adjusting the pressure, there could be a puncture or an issue with the TPMS sensor.

Understanding your dashboard warning lights is about more than just fixing immediate problems—it's about preventing bigger ones from cropping up later. These lights provide valuable insight into your car's health, often giving you the chance to address issues before they lead to costly repairs or leave you stranded.

COMMON SYMPTOMS AND THEIR CAUSES

Your car doesn't always rely on warning lights to let you know something is wrong. Often, it communicates through sounds, vibrations, leaks, or even unusual smells. By paying attention to these clues, you can diagnose issues early and decide whether they're something you can fix yourself or require professional help.

Unusual Noises and Vibrations: What They Mean

Every car makes noise, but not all noises are normal. Recognizing abnormal sounds or vibrations can help you catch problems before they escalate into expensive repairs.

- **Squealing or Screeching:** A high-pitched squeal when braking is a classic sign that your brake pads are worn and need replacing. Ignoring this can lead to rotor damage, which is significantly more expensive to fix. If the squealing happens while the engine is running, it's likely a slipping serpentine belt. Since this belt powers vital components like your alternator, power steering, and air conditioning, quickly replacing it is essential.

- **Knocking or Pinging:** Knocking sounds, often described as pinging, usually originate in the engine. They indicate an issue with the fuel-air mixture. Using low-octane fuel or having incorrect ignition timing can make the problem worse. Left unaddressed, this can lead to serious engine damage.
- **Grinding:** A grinding noise while braking is a clear signal that your brake pads are completely worn. If you don't act quickly, the rotors can be damaged, leading to costly repairs.
- **Vibrations:** Feeling vibrations while driving can point to issues like unbalanced tires, a bent wheel, or problems with the drivetrain. If the vibrations occur while braking, warped brake rotors are often to blame. This affects your stopping power and needs immediate attention.

Leaks, Smells, and Smoke: Clues from Fluids and Emissions

Your car's fluids and emissions provide another layer of insight into its health. Understanding what these signs mean can help you pinpoint problems faster.

Leaks: What's Dripping?

- **Oil Leaks:** Dark brown or black spots under your car often signal an oil leak. Common sources include the oil pan gasket or valve cover gasket. Promptly addressing leaks can prevent engine damage.
- **Coolant Leaks:** Bright green, orange, or pink puddles typically mean your car is losing coolant. Check around the radiator, hoses, or water pump for leaks. Driving with low coolant can lead to overheating, which can severely damage your engine.
- **Transmission Fluid Leaks:** A pool of red or brown fluid under your car often signals a transmission fluid leak. Since low fluid can damage the gearbox, it's important to address this issue quickly.

Smells: What's in the Air?

- **Burning Rubber:** This smell usually points to a slipping belt or an overheated tire. Inspect both to prevent further damage.
- **Rotten Eggs:** A sulfur-like smell often indicates a failing catalytic converter, which can affect your car's emissions system.
- **Gasoline:** The smell of gas typically signals a fuel leak—a serious safety hazard. Don't drive the car until the leak is repaired.

Smoke: Colors That Tell a Story

- **Blue Smoke:** Blue smoke from the exhaust means oil is burning, likely due to worn piston rings or valve seals. This requires immediate attention to prevent further engine damage.
- **White Smoke:** White smoke suggests coolant is burning, often due to a blown head gasket. This issue can cause significant engine problems if left unaddressed.

- **Black Smoke:** Black smoke signals an overly rich fuel mixture, often caused by a clogged air filter or faulty fuel injector. Replacing the filter or injector typically resolves the issue.

When you notice an unusual symptom, the first step is to gather as much information as possible. Does the noise occur only when you accelerate? Does the smell happen right after starting the car? These details are valuable for pinpointing the issue. For greater accuracy, consider using an OBD-II scanner to check for error codes. This handy tool reads diagnostic trouble codes (DTCs) from your car's computer, giving you a clearer picture of what's wrong. While it won't fix the problem for you, it's an invaluable resource for identifying the root cause.

Addressing symptoms early can save you time, money, and stress. For example, replacing brake pads when you hear squealing is far cheaper than replacing damaged rotors. Similarly, spotting a coolant leak early can prevent overheating and avoid costly engine repairs. By learning to interpret your car's sounds, vibrations, and other signals, you'll become a more confident and proactive car owner.

USING AN OBD-II SCANNER TO DECODE ERROR CODES

An OBD-II scanner is one of the most practical tools you can own for diagnosing car problems. This compact device plugs into your car's onboard diagnostic (OBD) port to retrieve error codes stored in its computer. Think of these codes as your car's way of giving you clues about what's wrong, whether it's a misfiring engine, a faulty sensor, or a loose gas cap. Let's break down how to use an OBD-II scanner and interpret some of the most common error codes.

How to Use an OBD-II Scanner

If you've never used an OBD-II scanner before, don't worry—it's easier than it sounds. Here's how to get started:

1. **Locate the Port:** Look under the dashboard near the driver's seat. You're searching for a rectangular connector—this is the OBD-II port.
2. **Plug It In:** Insert the scanner into the port and turn the ignition to the "on" position without starting the engine. This powers the scanner and lets it communicate with the car's computer.
3. **Retrieve Codes:** Follow the prompts on your scanner to extract the error codes. Many scanners will also display descriptions of the codes on the screen.
4. **Interpret the Codes:** Use the scanner's built-in database, your car's service manual, or online resources to understand what the codes mean. Some scanners even provide troubleshooting tips to guide you.

Once you've retrieved the codes, you're ready to dive into what they're telling you about your car.

25 Common Error Codes and What They Mean

Here's a guide to some of the most frequently encountered OBD-II error codes and what they indicate:

1. **P0171 (System Too Lean):** Too much air, not enough fuel—often due to a vacuum leak or faulty fuel injectors.
2. **P0174 (System Too Lean – Bank 2):** Similar to P0171 but specific to one side of the engine.
3. **P0300 (Random Misfire):** Misfires in multiple engine cylinders, typically from ignition or fuel delivery issues.
4. **P0301-P0308 (Cylinder Misfire):** Identifies misfires in specific cylinders (e.g., P0302 = Cylinder 2).
5. **P0420 (Catalyst System Efficiency Below Threshold):** Usually indicates a catalytic converter issue but can also be triggered by faulty upstream oxygen sensors. Proper diagnosis is key.
6. **P0440 (Evaporative Emission System Malfunction):** A problem in the fuel vapor system, such as a loose gas cap or faulty valve.
7. **P0455 (Evaporative Emission System Leak – Large):** Points to a significant fuel vapor leak.
8. **P0456 (Evaporative Emission System Leak – Small):** A smaller leak, often caused by a damaged or loose gas cap.
9. **P0101 (Mass Air Flow Sensor Circuit Range/Performance):** An issue with the sensor that measures air intake.
10. **P0113 (Intake Air Temperature Sensor Circuit High Input):** Indicates a problem with the air intake temperature sensor.
11. **P0128 (Coolant Thermostat Below Regulating Temperature):** The engine isn't warming up properly, often due to a stuck thermostat.
12. **P0141 (Oxygen Sensor Heater Circuit Malfunction):** A failing oxygen sensor can cause this code, affecting emissions.
13. **P0218 (Transmission Over Temperature):** Transmission fluid is overheating—needs immediate attention.
14. **P0221 (Throttle/Pedal Position Sensor Range/Performance):** An issue with the throttle or pedal sensor.
15. **P0335 (Crankshaft Position Sensor Circuit):** Indicates a problem with the crankshaft sensor, which impacts engine timing.
16. **P0401 (EGR Flow Insufficient):** The exhaust gas recirculation (EGR) valve may be clogged or malfunctioning.
17. **P0430 (Catalyst Efficiency Below Threshold – Bank 2):** Typically indicates an issue with the catalytic converter on Bank 2 but can also be caused by faulty upstream oxygen sensors. Accurate diagnosis is essential.
18. **P0500 (Vehicle Speed Sensor Malfunction):** Affects speedometer accuracy.

19. **P0550 (Power Steering Pressure Sensor Circuit):** A power steering sensor issue potentially affecting handling.
20. **P0601 (Internal Control Module Memory Check Sum Error):** Indicates an internal problem with the engine control module (ECM).
21. **P0700 (Transmission Control System Malfunction):** Requires further diagnosis of the transmission.
22. **P0720 (Output Speed Sensor Circuit Malfunction):** Affects the transmission's speed sensor.
23. **P1000 (OBD-II Readiness Test Not Complete):** Indicates the system hasn't completed its diagnostic tests after being reset.
24. **P1135 (Heated Oxygen Sensor Heater Circuit Malfunction):** A problem with the oxygen sensor's heating element, essential for emissions control.
25. **P0305 (Cylinder 5 Misfire):** Specifically points to a misfire in Cylinder 5.

While this list covers many of the most common OBD-II codes, there are hundreds more that your car could display. To find a complete database of error codes:

- **Check your OBD-II scanner's built-in library.** Many scanners come with an extensive code reference guide.
- **Explore online databases.** Websites like **OBD-Codes.com** and **OBD-II Trouble Codes.com** provide detailed explanations, possible causes, and troubleshooting tips for virtually every code.
- **Consult your car's service manual.** Manufacturer-specific codes (known as enhanced or manufacturer codes) are best interpreted using the manual for your vehicle.

These resources make it easy to expand your understanding of error codes and help you diagnose problems with confidence.

Acting on Error Codes: What's Next?

Once you've identified an error code, the big question is: What's the next step? Some issues, like a loose gas cap (P0456), are quick and easy to fix on your own. Others, such as a failing catalytic converter (P0420), might require the expertise of a professional. Here's how to navigate your options once you know what the code means.

- **Step 1--Research the Code:** Start by looking up the error code. Online forums, instructional videos, and even your car's service manual are great resources. These can provide insights into the most common causes and potential fixes for your specific error code. For instance, if you're dealing with a mass air flow sensor issue (P0101), you might learn that cleaning or replacing the sensor could resolve it.
- **Step 2--Assess Your Skills:** Next, consider whether the repair is something you can handle. Ask yourself a few questions: Do you have the necessary tools? Are you familiar with the part or system involved?

Simple fixes, like tightening a gas cap or replacing an air filter, are perfect for DIY. However, more complex issues, such as addressing a transmission problem, may require advanced skills and equipment.

- **Step 3--Know When to Call a Pro:** Not every problem is DIY-friendly, and that's okay. For complex repairs or issues involving critical systems like the transmission or engine, it's best to consult a trusted mechanic. Bringing your car in with a clear understanding of the error code and its possible causes not only helps you communicate effectively but also ensures you're less likely to pay for unnecessary work.

Understanding how to decode error codes puts the power back in your hands. It helps you understand what's wrong, avoid unnecessary repairs, and communicate effectively with a mechanic if needed. With an OBD-II scanner and a little knowledge, you can diagnose problems like a pro, saving yourself time, money, and stress. It's one of the simplest ways to stay in control of your car's health.

QUICK DIAGNOSTIC TECHNIQUES FOR BEGINNERS

Diagnosing car issues might sound intimidating, but it doesn't have to be. With a few simple techniques, you can identify problems before they escalate, saving you time, money, and unnecessary stress. Here's how you can start troubleshooting your car like a pro.

Listen and Observe

Your car has its own way of letting you know something's wrong—it's all about paying attention. Start by listening for unusual noises. Squealing might mean worn brake pads or a slipping belt. Grinding sounds could indicate brake or transmission issues while knocking or pinging noises might point to engine trouble.

Don't ignore strange smells, either. Burning rubber often signals a slipping belt or overheated tire, and the smell of gasoline could mean a fuel system leak, which is a serious safety hazard. Lastly, notice any changes in how your car feels, like unusual vibrations or sluggish acceleration. These subtle clues are often the first signs of a problem.

Perform a Visual Inspection

Sometimes the problem is hiding in plain sight. Pop the hood and take a good look around. Check for loose or damaged belts, cracked hoses, or signs of fluid leaks. If you notice any puddles under the car, note the color and texture—it could indicate anything from oil or coolant leaks to transmission fluid issues.

Don't forget to inspect your tires, too. Uneven wear, bulges, or debris embedded in the tread can point to alignment or suspension problems. A quick visual scan can reveal a lot and help you zero in on potential issues before they worsen.

Check Fluid Levels

Fluids are the lifeblood of your car, and low or dirty fluids can create a host of problems. Use the dipsticks or reservoir markers to check levels for oil, coolant, brake fluid, and transmission fluid. Make sure they're within the recommended range and look clean—discolored or sludgy fluids may need to be replaced. Staying on top of fluid maintenance is one of the easiest ways to keep your car running smoothly.

Test the Battery

A dead battery is one of the most common reasons for a car not starting. The good news? Testing it is simple. Using a multimeter, connect the probes to the battery terminals when the car is off. A reading below 12.4 volts means the battery may need charging or replacement. Regularly checking your battery ensures you won't be caught off guard on a busy morning.

By listening to your car, performing basic checks, and staying alert to small changes, you can catch problems early and take action before they grow into costly repairs. These simple diagnostic steps give you the confidence to handle minor issues on your own or explain the problem clearly if you need professional help.

Even if you're new to car care, these techniques empower you to stay ahead of potential issues, putting you in control of your car's health and reliability.

RECAP

Your car has its own way of telling you when something's wrong, and in this chapter, you've learned how to listen. From interpreting dashboard warning lights to identifying leaks, strange smells, or odd noises, you now know how to spot trouble before it becomes expensive.

We also covered practical tools, like OBD-II scanners, that take the guesswork out of diagnosing problems. Whether it's a loose gas cap or a more complex issue like a misfiring engine, these steps empower you to act with confidence. You've seen how simple inspections and attention to detail can save time and money while keeping your car in top shape.

But diagnosing is just the beginning. In the next chapter, we'll shift gears to uncover practical DIY secrets that can help you tackle common maintenance tasks with ease. Let's take everything you've learned so far and turn it into action.

CHAPTER 5

SECRETS OF DIY MAINTENANCE

SIMPLE FIXES MECHANICS CHARGE HUNDREDS FOR ARE EASIER THAN YOU THINK

Did you know that some mechanics charge up to 300% more for tasks you could easily do yourself? For instance, replacing a cabin air filter—a 10-minute job—can cost over $80 at a shop, while the part itself is often under $20. That's just one of many simple maintenance tasks that you can handle on your own, saving hundreds of dollars a year.

This chapter is your guide to mastering those straightforward jobs that mechanics frequently overcharge for. We'll cover everything from changing your oil and replacing spark plugs to swapping out wiper blades and brake pads. These tasks don't require advanced skills or expensive tools—just a bit of confidence and the willingness to get your hands dirty.

You'll also learn how to spot unnecessary upsells at the shop, like premature fluid flushes or overpriced air filter replacements. By understanding your car's real needs, you can avoid being pressured into paying for services that aren't necessary.

DIY maintenance isn't just about saving money, though that's a huge bonus. It's about feeling empowered and in control of your car's care. Each task you complete builds your confidence and deepens your understanding of how your vehicle works. By the end of this chapter, you'll have the tools, knowledge, and mindset to handle routine maintenance like a pro—saving time, money, and a lot of stress along the way.

HOW TO SAVE MONEY BY DOING IT YOURSELF

There's a common misconception that maintaining your car is overly complicated and requires expensive tools or advanced skills. The truth? Many car maintenance tasks are surprisingly simple and can be done with basic tools and a little guidance. Whether you're just starting out or looking for ways to save money, learning to care for your car can be both practical and empowering. Let's explore why embracing DIY maintenance is worth your time.

Cost Savings

One of the most compelling reasons to take on DIY car maintenance is the money you save. Mechanics often charge a premium, not just for their labor but also by marking up the cost of parts by 20% or more. By sourcing parts yourself—whether online or at a local auto supply store—you can cut those costs significantly.

Take my friend Kevin, for example. He used to bring his car to the shop for everything, even minor tasks like replacing air filters or windshield wipers. Then came the day he was charged $350 for a brake pad replacement. Frustrated, he decided to give it a try himself. After watching a few online tutorials, he purchased quality brake pads for $40 and replaced them in an afternoon. Not only did he save over $300 on that one job, but he realized how simple some of these tasks could be. Now, he tackles other projects like oil changes and spark plug replacements. Kevin estimates he saves over $1,000 each year—and his car runs just as smoothly, if not better.

Immediate Results

One of the biggest perks of DIY maintenance is the convenience. When you need an oil change, there's no waiting in line at a service center or scheduling an appointment. Instead, you can get it done in about 30 minutes at home. You'll also know the job was done correctly because you did it yourself—no second-guessing whether the mechanic rushed through it.

Plus, there's something uniquely satisfying about rolling up your sleeves and completing a task on your own terms. That sense of accomplishment, combined with the money saved, makes DIY maintenance a win-win.

Empowerment

Each time you complete a maintenance task, you build confidence and learn more about how your car operates. Starting with simple jobs, like replacing wiper blades or topping off fluids, creates a strong foundation for tackling more complex tasks down the line.

Kevin's story doesn't end with brake pads. After gaining confidence from that first project, he noticed his car occasionally stalling. Instead of heading straight to the shop, he diagnosed the problem himself as worn spark plugs. For under $30, he replaced them and saved another $200 compared to what the shop had quoted him. Each success has made him more self-reliant and less dependent on mechanics for routine care.

Preventive Maintenance

DIY maintenance isn't just about saving money—it's also about preventing costly problems. By regularly inspecting belts, hoses, and fluid levels, you can catch minor wear and tear before it leads to a breakdown.

For instance, during one of Kevin's routine checks, he noticed a crack in his serpentine belt. Replacing it himself cost around $25. Had he ignored it, the belt could have snapped, leaving him stranded on the side of the road and facing hundreds of dollars in towing and emergency repair fees. Regular DIY maintenance gives you control over your car's health and helps you avoid the stress of unexpected issues.

Learning to perform basic car maintenance is more than just a cost-saving measure—it's an investment in your skills and confidence. Tasks that might seem daunting at first, like changing oil or replacing brake pads, become second nature with a little practice.

Take a page from Kevin's book: start small and work your way up. Replacing an air filter or topping off fluids is a great place to begin. From there, you can move on to slightly more involved tasks, like replacing spark plugs or cleaning battery terminals. Each job you complete brings a sense of pride and independence, and the savings add up quickly.

SIMPLE MAINTENANCE TASKS MECHANICS OVERCHARGE FOR

Let's face it—mechanics are running a business, and while some repairs genuinely require their expertise, many routine maintenance tasks don't. These simple, straightforward jobs often come with hefty shop fees, but the good news is that with a few basic tools and a little guidance, you can handle them yourself. Here are some common tasks that mechanics tend to overcharge for:

- Windshield Wiper Replacement
- Brake Fluid Replacement
- Fuel Filter Replacement
- Headlight Bulb Replacement
- Spark Plug Replacement
- Drive Belt Replacement
- Rear Brake Light Replacement
- PCV Valve Replacement
- Fuse Replacement

These tasks aren't just about saving money—they're about building confidence in maintaining your own car. Every task you complete helps you better understand how your vehicle works, making you less reliant on mechanics for everyday care.

A Costly Lesson in Overcharges

Take Sarah, for example. She's someone who learned the hard way how quickly maintenance costs can add up when you don't know what to question. Sarah took her car to the shop for a routine oil change, expecting to pay around $50. By the time she left, her bill had skyrocketed to over $400.

What happened? The mechanic convinced her that her car urgently needed additional services, including a windshield wiper replacement, a cabin air filter change, new spark plugs, and a brake fluid flush. Feeling unsure and wanting to play it safe, Sarah approved the work. What she didn't realize was that most of these tasks were unnecessary—or could have been done at home for a fraction of the cost.

For instance, she paid $80 for the cabin air filter replacement. Later, she discovered the part would've cost her just $15, and the entire process would have taken five minutes. The story was similar for the other services. Once Sarah realized how much she had overpaid, she felt frustrated and, frankly, taken advantage of.

But instead of staying stuck in frustration, Sarah decided to take control. She started with small, manageable tasks like replacing her wiper blades and checking fluid levels. Over time, she tackled bigger jobs, such as changing her spark plugs. Now, Sarah estimates she saves hundreds of dollars a year on maintenance and feels far more confident every time she works on her car.

Sarah's experience is a perfect example of why it's so important to understand your car's needs and avoid falling for unnecessary services. It's not that mechanics are inherently bad—they're running businesses, and upselling is part of their model. But by educating yourself and learning to handle simple maintenance tasks, you can take control of your car care and avoid spending money on things you don't need.

In Chapter 11, we'll guide you through these tasks step by step. You'll learn how to perform each one, why it's important, and how to do it safely and efficiently. By the end, you'll feel empowered to handle these jobs on your own, saving both time and money—and you'll never feel pressured to approve unnecessary shop services again. But before we get to that, let's briefly discuss some insider tips that'll help you avoid unnecessary services.

INSIDER TIPS TO AVOID UNNECESSARY SERVICES

Let's be honest: car maintenance can feel overwhelming, especially when mechanics recommend services you're unsure about. The good news? Many upsells at repair shops aren't necessary, and by understanding your car's needs, you can save both time and money. Let's explore some actionable tips to help you avoid unnecessary services and take control of your car's care.

- **Consult Your Owner's Manual:** Your car's owner's manual is more than just a booklet in the glovebox—it's your vehicle's personal guide. It includes detailed maintenance schedules and procedures tailored to your car. If a mechanic suggests a service that doesn't align with your manual, double-check before agreeing. Sticking to the manufacturer's recommendations prevents unnecessary wear and ensures your car runs smoothly.
- **Be Skeptical of Frequent Oil Change Recommendations:** The old 3,000-mile rule for oil changes no longer applies to most modern vehicles. With advanced engines and synthetic oils, many cars only need oil changes every 7,500–10,000 miles. If a shop insists on the outdated rule, refer to your manual to confirm the proper interval and save yourself the extra expense.
- **Question Transmission Fluid Change Suggestions:** Transmission fluid changes are often over-recommended. While some shops might suggest replacing it every 12,000 miles, most cars don't need this service until 60,000–80,000 miles. Stick to the timeline in your manual, and don't let a premature flush disrupt your car's performance.
- **Avoid Unnecessary Fuel Injector Cleanings:** Fuel injector cleanings are commonly suggested as routine maintenance, but they're rarely necessary for modern cars running on high-quality gasoline. Unless your car is experiencing rough idling or reduced fuel efficiency, skip this service and save the extra cash.
- **Be Cautious with Engine Flushes:** An engine flush might sound like a great way to clean your engine, but it's unnecessary if you're consistent with regular oil changes. Worse, it can dislodge debris in a way that harms your engine. Unless your car manufacturer specifically recommends it or you're addressing sludge buildup, avoid this upsell.
- **Inspect Air Filters Before Replacing:** Air filters are another popular upsell, but they don't always need replacing when suggested. If the filter is only slightly dirty, clean it with a quick tap or vacuum instead of buying a new one. Replace it only if it's heavily soiled or damaged to ensure optimal engine performance.
- **Understand Brake Pad Wear:** Replacing brake pads is essential, but some mechanics may push you to replace them earlier than needed. Ask for a measurement of the pad thickness and compare it to the manufacturer's guidelines. Many brake pads have built-in wear indicators that squeal when they need replacing—listen for those signs before agreeing to premature replacements.
- **Be Wary of Coolant Flushes:** Coolant doesn't degrade as quickly as some mechanics claim. Unless it's visibly dirty or hasn't been replaced in over five years, a flush is likely unnecessary. Always check your manual for the recommended replacement interval and stick to it.
- **Monitor Battery Health Independently:** Before agreeing to a new battery, have yours tested to confirm it needs replacing. Many auto parts stores offer free battery testing, which helps you determine if the issue is a failing battery or something simpler, like corroded terminals or low charge.

- **Assess Tire Replacement Necessity:** Tires are one of the more expensive replacements, so don't swap them out unless truly needed. Use a tread depth gauge or the penny test to check your tires. If the tread is above 2/32 of an inch and there's no visible damage, they're likely fine. Rotating your tires regularly also helps them wear evenly and last longer.
- **Be Cautious with Add-On Services:** Headlight restoration, windshield treatments, or alignment checks are often suggested as add-ons. While some of these services might be helpful, they're frequently overpriced. Evaluate whether you truly need them or if you can address the issue yourself with a DIY kit.
- **Seek Second Opinions:** If a mechanic recommends an unexpected or costly service, don't feel pressured to decide on the spot. Take the time to get a second opinion or consult your owner's manual. A trusted mechanic or online research can provide clarity and help you avoid unnecessary work.

By following these tips, you can confidently avoid common upsells and ensure your car gets only the services it truly needs. Staying informed and proactive not only saves you money but also reduces the risk of unnecessary wear or damage from unwarranted procedures.

AVOIDING COSTLY MISTAKES IN CAR MAINTENANCE

Taking care of your car is one of the most important things you can do to ensure it runs smoothly and lasts for years. But even with the best intentions, certain missteps can lead to expensive repairs and headaches. The good news? Many of these costly mistakes are entirely avoidable. Let's take a closer look at ten common car maintenance mistakes, why they're so damaging, and how to steer clear of them.

- **Ignoring the Check Engine Light:** Few things are as easy to ignore as that little glowing light on your dashboard. It's tempting to brush it off, especially if your car seems to be running fine. But that check engine light is your car's way of telling you something's wrong. Ignoring it can allow small issues—like a loose gas cap or a failing sensor—to snowball into much bigger, more expensive problems. Instead of avoiding it, address the warning promptly. Many auto parts stores offer free diagnostic scans to help you understand what's wrong without an immediate trip to the mechanic.
- **Neglecting Oil Changes:** Skipping oil changes is a surefire way to shorten your engine's lifespan. Oil lubricates your engine's moving parts and helps prevent overheating. Over time, it breaks down and loses its effectiveness, leading to increased friction and wear. The result? Expensive engine repairs or even replacement. Stick to the oil change schedule recommended in your owner's manual—typically every 5,000–7,500 miles for most vehicles using synthetic oil.
- **Overlooking Tire Maintenance:** Your tires are your car's direct connection to the road, and neglecting them can cost you big time. Failing to maintain proper tire pressure or skipping regular rotations leads to

uneven wear, reduced fuel efficiency, and a higher risk of blowouts. Take a few minutes each month to check your tire pressure with a gauge, and don't forget to rotate them every 5,000–7,000 miles to extend their lifespan and keep your ride smooth.

- **Delaying Brake Repairs:** Squeaking, grinding, or spongy brake pedals are all signs your brakes need attention. Delaying repairs can lead to more serious damage, like worn rotors, which are much more expensive to replace than brake pads. If you notice anything unusual about your brakes, get them inspected and fixed promptly. Your safety—and your wallet—depends on it.
- **Using Incorrect Fuel:** Using the wrong type of fuel might seem harmless, but it can wreak havoc on your engine and fuel system. Your car's owner's manual specifies the correct fuel grade for a reason. For instance, using regular gas in a car designed for premium fuel can cause knocking, reduced performance, and long-term engine damage. Always double-check before filling up.
- **Neglecting Fluid Levels:** Your car relies on several key fluids to function properly, including coolant, brake fluid, and transmission fluid. Letting any of these run low can lead to overheating, brake failure, or transmission issues—all costly problems to fix. Make it a habit to check your fluid levels regularly and top them off as needed. If you notice a leak, address it immediately to prevent further damage.
- **Ignoring Unfamiliar Noises:** Unusual sounds—whether it's a grinding, clunking, or high-pitched squeal—are your car's way of telling you something's wrong. While it's easy to ignore a noise that seems minor, addressing it early can save you from much bigger repairs later. For example, a grinding sound during braking could mean worn pads, which, if left unchecked, can damage your rotors.
- **Letting Your Car Overheat:** Driving an overheating car is one of the fastest ways to cause catastrophic engine damage. Whether it's due to a coolant leak, a failing water pump, or a broken thermostat, overheating can lead to a blown head gasket or even complete engine failure. If your temperature gauge spikes, pull over immediately, let the car cool, and check your coolant level. Never ignore the signs of overheating—it's a mistake you can't afford.
- **Disregarding Dashboard Warning Lights:** Dashboard lights are your car's built-in alert system, and ignoring them is a recipe for disaster. Whether it's a low tire pressure warning, a battery alert, or a brake system light, these indicators are there to protect you and your car. Don't wait for the issue to escalate—investigate the cause and address it before it leads to costly repairs.
- **Not Taking Care of Your Filters:** Oil and air filters might not seem like a big deal, but they play a crucial role in keeping your engine running efficiently. Dirty or clogged filters reduce fuel efficiency, hinder performance, and can even cause engine damage over time. Regularly inspect and replace filters as needed—your owner's manual will provide specific intervals. It's a small investment that pays off in the long run.

Avoiding these costly car maintenance mistakes isn't just about saving money—it's about keeping your car safe, reliable, and running at its best. By staying proactive and addressing issues early, you'll prevent minor problems from turning into major repairs. Take the time to familiarize yourself with your car's needs, follow the maintenance schedule in your owner's manual, and listen to what your vehicle is telling you.

RECAP

DIY maintenance puts you in the driver's seat—literally and figuratively—when it comes to taking care of your car. In this chapter, we covered some of the easiest and most overcharged tasks, like changing your oil, replacing air filters, and swapping out wiper blades. These aren't just manageable; they're budget-friendly, too. With a few basic tools and a little practice, you can handle these jobs without breaking a sweat—or the bank.

We also dug into how to avoid common upsells at repair shops. Mechanics often recommend premature services, like fluid flushes or unnecessary air filter replacements, which can quickly inflate your bill. By knowing your car's real needs and trusting your owner's manual, you can confidently say "no thanks" to extras you don't need.

But saving money isn't the only benefit. Every task you complete on your own gives you a deeper understanding of your car and builds your confidence. Those small wins add up, making you more self-reliant and less dependent on professional services for everyday maintenance.

Now that you've mastered the basics, it's time to shift gears. In the next chapter, we'll focus on two of the most essential systems in your car: brakes and suspension. These are the unsung heroes of a safe and comfortable ride, and with a bit of attention, you can keep them performing at their best.

CHAPTER 6

MASTERING THE ART OF BRAKE AND SUSPENSION CARE

STRONG BRAKES AND STURDY SUSPENSION ARE YOUR CAR'S SAFETY NET—KEEP THEM SHARP

A few years ago, my friend Lisa called me in a panic. Her car was making a terrible grinding noise every time she hit the brakes. She had ignored the squealing sound that started weeks earlier, thinking it wasn't a big deal. By the time she got to a mechanic, the brake pads were completely worn down, and the rotors were damaged. What should have been a $50 fix ended up costing her over $500.

Brakes and suspension are two of the most important systems in your car. They're not just about safety—they're what make your car comfortable to drive and responsive when it counts. Yet, like Lisa, many people miss the early warning signs that something's wrong, leading to expensive repairs that could have been avoided.

In this chapter, we'll break it all down. You'll learn how brakes work, how to tell when your pads or rotors need replacing, and what different noises or vibrations might mean. We'll also cover your suspension system, teaching you how to spot issues like worn shocks or uneven tire wear that can affect both safety and handling.

Even if you're new to car care, you'll find actionable steps here. From replacing brake pads to performing a bounce test on your suspension, we'll guide you through the basics so you can handle these tasks with confidence. These aren't just skills to save you money—they're what keep you safe and your car performing its best. Let's dive in and ensure your car is ready for whatever the road throws your way.

WHAT YOU NEED TO KNOW ABOUT THE BRAKE SYSTEM

Your car's brakes are one of the most critical systems for your safety. Understanding how they work and knowing when they need attention can save you money and give you peace of mind on the road. Let's break down the essentials of your braking system and how to maintain it effectively.

Brakes are all about converting kinetic energy into heat energy to bring your car to a stop. Here's the gist: when you press the brake pedal, hydraulic fluid travels through brake lines to the calipers. These calipers press the brake pads against the rotors—large discs attached to your wheels. The friction this creates is what slows down and stops your car.

Modern braking systems take it a step further. Features like anti-lock braking systems (ABS) prevent wheel lockup during sudden stops, and electronic brakeforce distribution (EBD) balances braking power across all four wheels. Understanding these components helps you recognize potential issues and address them before they escalate.

Signs Your Brakes Need Attention

Your brakes are one of the most important safety features in your car, and thankfully, they rarely fail without warning. Catching the signs early not only helps you avoid costly repairs but also ensures your safety on the road. Let's look at the key indicators that your brakes may need attention.

- **Squealing or Squeaking Noises:** That high-pitched squeal when you press the brake pedal isn't just annoying—it's your car's way of saying, "Hey, my brake pads are wearing out!" Many brake pads come with built-in wear indicators that make this sound to alert you. While it's not an emergency, it's a clear signal that you should plan to replace your brake pads soon.
- **Grinding Noise:** If you hear grinding, it's time to act fast. This noise means your brake pads have worn down completely, allowing the metal backing to scrape against the rotors. Ignoring this can cause significant damage, turning a simple pad replacement into a costly rotor repair or replacement.
- **Vibrations While Braking:** Do you feel your steering wheel or brake pedal vibrating when you apply the brakes? This often points to warped rotors, which can happen if the rotors overheat from heavy braking. Warped rotors compromise braking efficiency and need to be resurfaced or replaced to restore smooth stopping power.
- **Longer Stopping Distances:** If it's taking longer to stop than it used to, that's a serious problem. Thin brake pads, contaminated brake fluid, or worn-out rotors could all be culprits. Whatever the cause, delayed stopping distances put you and others on the road at risk, so don't delay addressing the issue.

- **Brake Warning Light:** Modern cars are equipped with dashboard lights to alert you to braking system problems. If the brake warning light comes on, don't ignore it. It could indicate anything from low brake fluid to an issue with the anti-lock braking system (ABS). Either way, it's a sign that your brakes need immediate attention.

Ignoring brake warning signs isn't just dangerous—it's costly. A minor issue like worn brake pads is far less expensive to fix than the damage caused by neglecting them. For instance, allowing grinding brakes to go unchecked can ruin your rotors, adding hundreds of dollars to your repair bill.

But it's not just about money. Reliable brakes are essential for avoiding sudden obstacles, navigating heavy traffic, and ensuring your car stops when you need it to. Addressing brake problems promptly gives you the confidence that your vehicle is ready for anything the road might throw at you.

You don't have to be a professional to check your brakes. A quick at-home inspection can help you catch potential issues before they escalate:

- **Visual Check**: Peek through your wheel spokes to see the brake pads. If they're less than 1/4 inch thick, it's time for a replacement.
- **Listen for Sounds**: Keep an ear out for squealing, grinding, or clicking noises while driving. These sounds often indicate the need for maintenance.
- **Feel for Vibrations**: Pay attention to any unusual vibrations in the brake pedal or steering wheel when braking.

If something doesn't feel or sound right, it's better to address it now than to deal with bigger, more expensive problems later.

Components Breakdown: Pads, Rotors, and Calipers

Understanding the core components of your braking system helps you maintain it effectively and identify potential issues before they escalate. Let's take a closer look at the three main parts that make up your brakes: pads, rotors, and calipers.

Brake Pads: The Frontline of Friction

Brake pads are the heroes of your braking system. Every time you press the brake pedal, the pads clamp down on the rotors to create friction that slows or stops your car. But they're designed to wear out over time, so regular replacement is part of good brake care. There are three main types of brake pads:

- **Ceramic Pads**: Quiet and long-lasting, these are great for everyday driving but come with a higher price tag.

- **Metallic Pads**: Durable and effective, metallic pads are a solid choice for heavier vehicles or performance driving, though they can be a bit noisier.
- **Organic Pads**: The most budget-friendly option, organic pads wear out faster and may need more frequent replacements.

Remember, if you hear squealing or squeaking noises when braking, your pads are likely nearing the end of their life. If you hear grinding, act immediately—that means the pads are completely worn, and the metal backing is scraping against the rotors.

Rotors: The Surface for Stopping Power

Rotors are the metal discs that the brake pads press against to generate the friction that stops your car. While they're built to handle high heat and pressure, they aren't immune to wear and tear over time. Rotors come in a few different styles:

- **Smooth Rotors**: Ideal for standard driving and the go-to choice for most vehicles.
- **Slotted Rotors**: Designed with grooves to dissipate heat, making them better suited for heavy-duty or high-performance driving.
- **Drilled Rotors**: Perforated for improved cooling, these are often found in sports or high-performance cars.

Warped rotors can cause vibrations in the brake pedal or steering wheel during braking. Deep grooves or visible damage indicate it's time to resurface or replace them.

Calipers: The Hydraulic Clamps

Calipers are the hydraulic components that house your brake pads and push them against the rotors. Think of them as the clamps that make braking possible. Over time, calipers can develop two common issues:

- **Seizing**: When a caliper seizes, it doesn't release the pads properly, causing uneven wear or constant drag on the rotors.
- **Leaking**: A brake fluid leak in the calipers reduces hydraulic pressure, compromising your braking power.

Keep in mind: if your car pulls to one side when braking or if the brakes feel weak, it could point to a caliper problem. Ultimately, your brake pads, rotors, and calipers are a finely tuned team. The calipers apply hydraulic pressure to push the brake pads against the rotors, creating the friction needed to stop your car. If any one of these components is worn or damaged, the entire system suffers, affecting your ability to brake safely.

Brake issues rarely resolve themselves. Catching wear and tear early is crucial for avoiding bigger problems—and bigger bills—down the road. A simple visual inspection can reveal worn pads, warped rotors, or leaking calipers.

Don't forget to listen for unusual noises and pay attention to vibrations in the brake pedal or steering wheel, as these are often the first signs something's off.

DIY Tips for Brake Pad Replacement

Replacing brake pads is a beginner-friendly task that can save you hundreds of dollars. Here's how to do it safely:

1. **Gather Your Tools**: You'll need a car jack, lug wrench, socket set, C-clamp, and new brake pads. Don't forget safety goggles and gloves.
2. **Ensure Safety**: Park on a level surface, engage the parking brake, and use wheel chocks to prevent movement.
3. **Remove the Wheel**: Loosen the lug nuts, lift the car with the jack, and support it with jack stands. Remove the wheel to access the brakes.
4. **Access the Caliper**: Unscrew the caliper bolts and slide the caliper off the rotor. Use a rope or hanger to support the caliper so it doesn't dangle by the brake line.
5. **Replace the Brake Pads**: Remove the old pads and inspect the caliper piston. Use a C-clamp to compress the piston and create space for the new pads. Install the new pads, ensuring they're seated properly.
6. **Reassemble the Brakes**: Reattach the caliper, tighten the bolts, and mount the wheel. Tighten the lug nuts and lower the car.
7. **Test Your Brakes**: Pump the brake pedal several times to restore pressure. Test them in a safe, low-speed area before hitting the road.

Taking the time to inspect and replace components like brake pads will extend your car's lifespan and provide a smoother, safer ride. In the next section, we'll focus specifically on suspension care, including how to inspect and replace shocks, struts, and other critical components.

UNDERSTANDING YOUR CAR'S SUSPENSION SYSTEM

Your car's suspension system is about much more than providing a smooth ride. It's a key player in your vehicle's safety and performance, ensuring stability, control, and comfort on any road. Understanding how it works—and knowing what to watch for—helps you stay ahead of potential problems while avoiding costly repairs.

Think of your suspension system as your car's shock absorber for the road. It connects the car's frame to its wheels and cushions the impact of bumps, dips, and uneven surfaces. Beyond comfort, the suspension system keeps your tires in contact with the ground, providing the stability and control you need for safe braking, cornering, and acceleration.

A well-maintained suspension system ensures predictable handling and gives you confidence behind the wheel. When it's working as it should, you might not even notice it. But when something's wrong, it can affect both your safety and driving experience.

Components of the Suspension System

To keep your suspension system in top shape, it's important to know its key parts and how they work together:

- **Shocks and Struts:** These components are the backbone of your suspension system. Shocks control the up-and-down motion of the car, while struts combine shock absorption with structural support. Together, they prevent excessive bouncing and keep your car stable.
- **Springs:** Springs—either coil or leaf—compress and expand to absorb road impacts. They work hand-in-hand with shocks and struts to handle uneven terrain.
- **Bushings:** Bushings are small, flexible connectors found at key points like control arms and sway bars. They reduce vibration and allow suspension parts to move smoothly.
- **Ball Joints and Tie Rods:** These components connect your suspension system to the steering system. Ball joints allow for smooth articulation, while tie rods ensure precise steering control.

Understanding these parts helps you recognize symptoms of wear and know where to focus your attention during maintenance.

Common Suspension Issues

Like any system in your car, the suspension experiences wear and tear over time. Here are some red flags to watch for:

- **Uneven Tire Wear**: If your tires are wearing unevenly or faster than usual, it could be a sign of misaligned suspension or worn components. Regular tire rotations and alignment checks can help prevent this issue.
- **Excessive Bouncing**: Does your car feel like it's still bouncing after hitting a bump? That's often a clear indicator that your shocks or struts are worn out.
- **Noises Over Bumps**: Squeaks, clunks, or rattles when driving over rough roads are typically caused by loose components, worn bushings, or damaged ball joints.
- **Pulling to One Side**: If your car drifts to one side while driving or braking, it may indicate a suspension misalignment or damaged parts like tie rods or control arms.

Addressing these issues early prevents further damage and ensures your car remains safe to drive.

HOW TO PERFORM A DIY SUSPENSION INSPECTION

You don't have to be a mechanic to spot suspension problems. Here's how to perform a quick DIY inspection:

1. **Bounce Test**: Push down firmly on the hood or trunk of your car and let go. If the car bounces more than twice, your shocks or struts may need replacing.
2. **Visual Inspection**: Look under your car for cracked bushings, leaking shocks, or worn springs. Also, check your tires for uneven wear, as it's often linked to suspension problems.
3. **Listen for Noises**: Take a short drive over a bumpy road and listen for clunks, squeaks, or rattles. These sounds usually indicate loose or worn components.

Spotting issues early can save you money and prevent more significant problems down the line.

Replacing Shocks and Struts: What to Expect

Replacing shocks and struts might seem daunting, but with the right tools and some preparation, it's a job you can tackle at home.

Tools You'll Need

- Jack and jack stands
- Lug wrench
- Socket set
- Torque wrench
- Spring compressor (for struts)

Steps for Replacement

1. **Ensure Safety**: Park on a level surface, engage the parking brake, and use wheel chocks.
2. **Lift the Car**: Loosen the lug nuts, lift the car with a jack, and secure it with jack stands.
3. **Remove Old Shocks or Struts**: Unscrew the bolts securing the shocks or struts and remove them. For struts, use a spring compressor to safely handle the spring.
4. **Install the New Parts**: Position the new shocks or struts and secure them with bolts. When tightening, be sure to consult your vehicle's manual for the precise torque values, as these can vary depending on the car model and brand. Using the correct torque ensures safety and prevents damage to the components.
5. **Test Drive**: After reassembling, take a short drive to ensure the new components are functioning properly.

Driving with worn suspension components isn't just uncomfortable—it's unsafe. Faulty shocks and struts reduce your car's ability to handle sharp turns, sudden stops, or uneven roads. Neglecting these issues can also cause premature wear on other parts, like your tires and brakes, leading to even higher repair costs.

Regular inspections and timely replacements keep your car safe, extend the lifespan of related components, and enhance your overall driving experience.

Caring for your suspension system might feel intimidating at first, but with regular checks and basic maintenance, it's manageable. By paying attention to warning signs like uneven tire wear or excessive bouncing, you can address problems early and save yourself from expensive repairs.

RECAP

In this chapter, we dug into the essentials of brake and suspension care—two systems that do more than just keep your car moving; they keep you safe. Understanding how brakes work, from converting energy into stopping power, gives you the tools to identify issues early. Whether it's squealing pads, a grinding noise, or the unsettling pull of your car to one side, you now know these aren't just inconveniences—they're warning signs.

We also talked about suspension care and how it impacts both safety and comfort. Uneven tire wear, excessive bouncing, or clunks over bumps aren't just nuisances; they're clues that something might be wrong with your shocks, struts, or other components. Armed with practical tips like how to replace brake pads or perform a simple bounce test, you're better equipped to maintain these systems and avoid costly repairs.

The best part? These tasks aren't as intimidating as they seem. With the right tools and a little guidance, you can tackle these jobs at home. Not only will you save money, but you'll also gain confidence in your ability to care for your car. Plus, knowing your brakes and suspension are in top shape means one less thing to worry about when you're behind the wheel.

Now that you've got a handle on these critical systems, it's time to look under the hood. In the next chapter, we'll explore the secrets to keeping your engine running like a dream. From routine maintenance to insider tips mechanics don't always share, you'll learn how to keep your car's powerhouse performing at its best.

CHAPTER 7

ENGINE MAINTENANCE SECRETS MECHANICS DON'T SHARE

A LITTLE DIY CARE FOR YOUR ENGINE SAVES THOUSANDS IN AVOIDABLE BREAKDOWNS

Did you know that a single worn spark plug can lower your engine's fuel efficiency by up to 30%? That small, often-overlooked part has a big job to do, and when it's not working right, your wallet feels it at the pump. Here's the good news: replacing spark plugs is one of the simplest and most affordable ways to keep your engine humming along, and you don't need a mechanic to get it done.

In this chapter, we'll demystify engine maintenance and share a few secrets that mechanics might not mention. You'll learn how to replace spark plugs and ignition wires—two key players in keeping your car starting smoothly and running efficiently. We'll also tackle the unsung heroes under your hood: serpentine and timing belts. These quiet workhorses keep everything in sync, and knowing when to maintain or replace them can save you from expensive repairs down the road.

But there's more to it than just parts. Cleaning your throttle body and mass airflow sensor may sound technical, but these tasks can dramatically improve your car's performance without a hefty price tag. Plus, we'll show you how to keep your radiator in top shape to avoid overheating—one of the leading causes of engine failure.

This isn't about turning you into a professional mechanic. It's about giving you the tools and confidence to take control of your car's care. By the end of this chapter, you'll know how to handle key maintenance tasks yourself, keeping your engine running smoothly and saving money along the way. Let's get started and make sure your car is ready to tackle every mile ahead!

SPARK PLUGS AND IGNITION WIRES: THE LIFEBLOOD OF IGNITION

Let's start with one of the simplest yet most impactful parts of your engine: spark plugs. These small but mighty components are essential for starting your car. Their job? To ignite the air-fuel mixture in your engine, creating the tiny explosions that power your vehicle. Without functioning spark plugs, you're going nowhere.

Recognizing When to Replace Spark Plugs

Like any consumable part, spark plugs wear out over time. Knowing the signs can help you replace them before they cause bigger issues. Sluggish starts, misfiring engines, poor fuel efficiency, or rough idling are all red flags. If your car feels less responsive or you're burning through gas faster than usual, it's worth investigating the spark plugs. A quick visual inspection can also reveal the condition of your plugs. Remove one and check the tip:

- **Blackened or Oily Tip**: Indicates incomplete combustion or an oil leak.
- **Excessive Wear**: Suggests it's time for a replacement.
- **Burnt Appearance**: Points to overheating or incorrect gapping.

The replacement interval for spark plugs varies significantly based on the material. Copper spark plugs typically last around 30,000 miles, while platinum or iridium plugs can last between 60,000 and 100,000 miles. Most cars fall within this range, but it's always best to consult your owner's manual for the specific recommendation for your vehicle.

DIY Spark Plug Replacement

It's a common myth that spark plug replacement is always a job for professionals. In reality, it's one of the easiest DIY maintenance tasks, even for beginners. As long as you have the right tools and follow the steps carefully, there's no reason you can't handle it yourself and save money.

What You'll Need:

- Socket wrench
- Spark plug socket
- Gap gauge

How to Do It:

1. **Locate the Spark Plugs**: Your manual will guide you to their location, often under a plastic engine cover.
2. **Remove the Old Plugs**: Use the spark plug socket to unscrew them. Be gentle to avoid damaging the threads.
3. **Check the Gap**: Ensure the new spark plugs match the manufacturer's gap specifications using a gap gauge. Proper gapping is crucial for optimal performance.
4. **Install the New Plugs**: Screw them in by hand first to prevent cross-threading. Tighten them with a wrench, but avoid overtightening.
5. **Reconnect the Wires**: If your car uses ignition wires, make sure they're securely reattached.

Another misconception is that all spark plugs are the same. In truth, the type of spark plug you choose—ceramic, iridium, or copper—can significantly affect your car's performance. For example, iridium plugs last longer and offer better efficiency but come with a higher price tag. Weigh the upfront cost against longevity to decide what's best for your vehicle.

Ignition Wires: Don't Overlook Them

While spark plugs get most of the spotlight, ignition wires play an equally important role. These wires deliver the electrical charge from the ignition coil to the spark plugs, igniting the air-fuel mixture.

Over time, ignition wires can crack, fray, or corrode, leading to weak or inconsistent sparks. This can cause misfires, reduced performance, or even engine damage. If you notice visible wear or experience symptoms like misfiring or rough idling, it's time to replace the wires.

- **Pro Tip**: Replace one wire at a time to avoid mixing up the connections. Some modern vehicles use ignition coil packs instead of wires, which require a slightly different approach. Check your manual for guidance.

Spark plugs and ignition wires are small parts that make a big difference in your engine's health. Replacing them when they're worn improves fuel efficiency, reduces emissions, and keeps your car running smoothly. Plus, learning to handle these replacements yourself is a great way to save money and take control of your car care routine.

SERPENTINE AND TIMING BELTS: WHEN AND HOW TO REPLACE THEM

When it comes to car maintenance, belts don't get nearly as much attention as components like brakes or tires, yet they play a crucial role in keeping your engine running smoothly. Belts quietly ensure everything stays in sync under the hood. The serpentine belt powers critical accessories like the alternator, power steering pump, and air conditioning. Meanwhile, the timing belt ensures your engine's valves open and close at precisely the right time, keeping the combustion process running efficiently.

Think of belts as the "connective tissue" of your engine. If one fails, it can take down essential systems like power steering, your alternator, or even your entire engine. Replacing a serpentine belt before it snaps can prevent you from being stranded on the side of the road while keeping up with timing belt maintenance can save you thousands in repairs.

Despite their importance, belts are often overlooked until something goes wrong. But here's the thing: worn-out belts give you warning signs if you know what to look (or listen) for. Catching these signs early can save you from costly repairs or unexpected breakdowns.

Recognizing the Signs of Worn Belts

So, how do you know when a belt is nearing the end of its life? Here are some common indicators:

- **Cracking**: Look for fine lines or cracks along the belt's surface. Over time, heat and wear cause the belt material to deteriorate.
- **Glazing**: A shiny or glossy appearance often means the belt is slipping, reducing its ability to grip the pulleys.
- **Loss of Tension**: If the belt feels loose, it may no longer be keeping connected components—like the alternator or power steering—functioning properly.
- **Squealing Sounds**: This is one of the most obvious signs. If you hear a high-pitched squeal coming from the engine bay, it's likely the serpentine belt is slipping or losing tension.

Ignoring these signs can lead to bigger headaches down the road, like losing your power steering or alternator function mid-drive.

How to Replace a Serpentine Belt

The good news? Replacing a serpentine belt is one of the easier maintenance tasks you can tackle yourself. With a bit of preparation and patience, you'll have it swapped out in no time. Here's how:

1. **Locate the Belt**: Start by consulting your owner's manual. Most modern cars have a belt routing diagram under the hood, making this step simple.
2. **Release Tension**: Use a wrench or serpentine belt tool to rotate the tensioner pulley. This will loosen the belt, allowing you to remove it.
3. **Remove the Old Belt**: Carefully slide the old belt off the pulleys. Pay attention to its routing so you can install the new one correctly.

4. **Install the New Belt**: Follow the diagram to route the new belt over the pulleys. Make sure it sits snugly in the grooves.
5. **Reapply Tension**: Slowly release the tensioner pulley, so it applies pressure to the new belt. Double-check that the belt is properly aligned.

While replacing the serpentine belt, inspect the pulleys for wear or rough edges, which can damage the new belt. Regularly inspecting the belt during oil changes or routine maintenance will also help you catch issues early.

How to Replace the Timing Belt

Unlike the serpentine belt, the timing belt is a bit trickier to replace and often requires professional help. Why? It's buried deep inside the engine, and replacing it involves precision. If the timing belt isn't installed perfectly, it can cause severe engine damage.

Most timing belts need to be replaced every 60,000 to 100,000 miles, depending on your vehicle's manufacturer's recommendations. Don't ignore this schedule—if a timing belt fails, it can cause your engine's valves and pistons to collide, leading to catastrophic (and expensive) repairs.

Symptoms of a Worn Timing Belt

While it's not as easy to inspect as a serpentine belt, the timing belt does give you clues when it's starting to fail:

- **Ticking Noise**: A rhythmic ticking sound coming from the engine can indicate a loose or worn belt.
- **Engine Misfires**: If the belt is slipping or worn, it can throw off the engine's timing, causing misfires or rough running.
- **Oil Leaks**: If you notice oil leaks near the timing belt cover, this could mean oil is compromising the belt's integrity.

Even if you're not replacing the timing belt yourself, understanding its role and recognizing these symptoms will help you have a more informed conversation with your mechanic. Don't be afraid to ask questions or request to see the worn belt if the shop recommends a replacement.

Belts may not seem flashy, but they're vital for your car's operation. Regular inspections and timely replacements will keep your engine happy and your wallet safe from unexpected expenses.

CLEANING YOUR THROTTLE BODY AND MASS AIRFLOW SENSOR

The throttle body and mass airflow (MAF) sensor may not get much attention, but they're crucial to your car's performance. These parts regulate the airflow entering your engine, ensuring the air-fuel mixture is just right for combustion. When clean, they help your engine run efficiently and provide smoother acceleration. Let them get dirty, though, and you'll likely face rough idling, stalling, or sluggish performance.

Imagine trying to breathe through a straw that's full of gunk. It wouldn't be easy, right? That's exactly how your engine feels when the throttle body or MAF sensor gets dirty. Over time, grime, carbon buildup, and dust accumulate, restricting airflow and leading to poor fuel economy, weak acceleration, and even stalling.

Unfortunately, mechanics sometimes use this opportunity to upsell expensive "fuel system cleanings" or convince you to replace parts unnecessarily. The truth? A good cleaning is often all these components need, and the best part is—you can do it yourself at home in under an hour. Cleaning these parts isn't as complicated as it sounds, and with minimal tools and effort, you'll keep your engine breathing easy and running its best.

How to Clean Your Throttle Body

The throttle body is responsible for regulating airflow to your engine. When you press the gas pedal, the throttle plate inside opens to allow more air in. Carbon buildup around the plate can cause it to stick, reducing responsiveness. Here's how to clean it:

- **Access the Throttle Body:** Open your car's hood and locate the air intake hose (refer to your owner's manual for guidance). Disconnect the intake hose where it connects to the throttle body. You'll see a round metal housing with a butterfly-shaped throttle plate inside.
- **Inspect for Dirt and Grime:** Look inside the throttle body for black soot or buildup. Don't worry—it's a common issue.
- **Clean with Throttle Body Cleaner:** Use a throttle body cleaner (available at auto stores) and a clean, lint-free cloth. Spray the cleaner onto the cloth instead of directly into the throttle body to avoid damaging sensors. Wipe away the grime, focusing on the edges of the throttle plate.
- **Move the Throttle Plate:** Gently open the throttle plate to access hidden areas and clean thoroughly.
- **Reassemble and Test:** Reconnect the air intake hose and double-check all connections. Start the car and let it idle for a few minutes to recalibrate the system. You should notice smoother idling and better throttle response.

How to Clean Your Mass Airflow (MAF) Sensor

The MAF sensor measures the air entering your engine, helping the car adjust fuel delivery. A dirty MAF sensor can send incorrect readings, leading to poor fuel efficiency and performance. Cleaning it is simple:

- **Locate the Sensor:** Find the MAF sensor near the air filter housing. It's a small rectangular piece with an electrical connector.
- **Remove It Carefully:** Disconnect the electrical plug and unscrew the sensor. Handle it gently—it's delicate and costly to replace.
- **Clean with MAF Sensor Cleaner:** Only use a designated MAF sensor cleaner. Avoid carburetor cleaner, compressed air, or throttle body cleaner—they can damage the sensor's fragile wires. Spray the cleaner directly onto the sensor's wires and components. Don't touch the wires with your hands or tools.
- **Let It Dry Completely:** Allow the sensor to air dry fully for at least 10–15 minutes before reinstalling it.
- **Reinstall and Test:** Reattach the sensor and plug it into the electrical connector. Start your engine and check for smoother idling and improved performance.

Keeping your throttle body and MAF sensor clean starts with a few simple habits that can prevent buildup and keep your engine running smoothly. First, check your air filter regularly. A dirty or clogged air filter allows dust and debris to bypass it, settling on the throttle body and MAF sensor. Replacing your air filter when it's dirty ensures clean airflow and helps prevent grime buildup on these critical components.

Second, use quality fuel whenever possible. Premium fuel often contains detergents that help reduce carbon buildup over time. It's a small investment that can go a long way in keeping your engine efficient and reducing the need for frequent cleanings.

Finally, make cleaning these components part of your annual maintenance routine. If you drive in dusty or harsh conditions, consider cleaning them more often to keep everything performing at its best.

One big myth is that cleaning the throttle body or MAF sensor is a job for professionals. Mechanics might claim these parts are too delicate for DIY cleaning, but the truth is, with the right cleaner and a careful approach, you can handle it yourself without any trouble.

Another common misconception? That these parts always need replacing when your engine starts acting up. Often, a thorough cleaning is all it takes to restore their function and save you hundreds of dollars. Before committing to expensive replacements, give cleaning a shot—you'll be surprised how much of a difference it can make in your car's performance.

Overall, cleaning your throttle body and MAF sensor might sound intimidating, but it's easier than you think. Taking the time to tackle this job at home saves you money, keeps your engine running efficiently, and avoids unnecessary repairs.

KEEPING YOUR ENGINE COOL: RADIATOR MAINTENANCE

Your engine works hard, generating intense heat every time you drive. Keeping that heat under control is the radiator's main job, and it's a role you don't want to take for granted. Neglecting your cooling system can lead to overheating, which is one of the fastest ways to damage an engine. Fortunately, maintaining your radiator and cooling system is easier than you might think—and it can save you from costly repairs down the road. But before we dive into that, let's quickly chat about why radiator maintenance is essential.

Think of the radiator as your engine's personal cooling center. It circulates coolant to absorb heat from the engine and then releases that heat into the air through its fins. If the radiator becomes clogged, runs low on coolant, or develops a leak, your engine can overheat quickly. Overheating isn't just inconvenient—it can warp the engine block, damage gaskets, and lead to repairs that cost thousands of dollars. That's why a little maintenance goes a long way.

Checking Coolant Levels

Coolant, also called antifreeze, prevents your engine from overheating in summer and freezing in winter. Regularly checking your coolant levels is quick, simple, and one of the easiest ways to catch small issues early. Here's how to do it safely:

- **Wait for the Engine to Cool:** Never open the radiator cap while the engine is hot. The coolant is under pressure and can cause serious burns.
- **Check the Levels:** Look at the reservoir tank—it has "min" and "max" lines. Your coolant should always sit between those markings. If it's low, you may have a leak or another problem.
- **Inspect the Color:** Healthy coolant is bright green, orange, or pink (depending on the type). If it looks brown, murky, or rust-colored, it's time for a coolant flush.

Using a coolant tester can help you confirm that the mix is still effective. This simple habit can save you from major headaches down the line.

Flushing the Radiator

Coolant degrades over time, losing its ability to protect against heat and rust. Plus, debris and rust particles can contaminate the system, leading to clogs. That's why flushing the radiator every two years or 30,000 miles is so important. Here's how to do it:

1. **Drain the Old Coolant:** Place a container under the radiator, open the drain valve, and let the coolant flow out. Be sure to dispose of it safely—it's toxic.

2. **Flush with Water:** Close the valve, fill the radiator with distilled water, and run the engine for about 10 minutes. Distilled water prevents introducing minerals that could lead to corrosion, which can cause extensive damage to the cooling system. Let the engine cool, then drain the water.
3. **Refill with Fresh Coolant:** Use the correct coolant type for your car (check the owner's manual) and mix it with distilled water if needed. Fill the radiator and the reservoir to the proper levels.
4. **Bleed the System:** Run the engine with the radiator cap off to let air bubbles escape. Once the bubbles stop, replace the cap.

Regular flushing keeps your radiator clean, prevents corrosion, and ensures your cooling system works efficiently.

Inspecting Hoses and Connections

Hoses are the pathways that move coolant between the radiator, engine, and other components. Over time, they can crack, harden, or loosen, which can lead to leaks. Here's what to watch for:

- **Feel for Flexibility:** Squeeze the hoses—if they feel brittle, stiff, or unusually soft, it's time to replace them.
- **Check the Connections:** Ensure clamps are secure but not overly tight, as overtightening can damage the hose.
- **Look for Leaks:** Check for puddles under your car or dried coolant residue near hose connections. Even small leaks can cause big problems if ignored.

Avoiding Overheating: What to Do

Even with proper maintenance, things can occasionally go wrong. If you see the temperature gauge creeping toward the red zone, don't ignore it. Here's what to do:

1. **Pull Over Safely:** Turn off the engine immediately to prevent further damage.
2. **Let It Cool:** Wait at least 20–30 minutes before popping the hood.
3. **Check Coolant Levels:** Once it's cool, check the reservoir. If it's low, add coolant or water to get to a safe level.
4. **Call for Help:** If the problem persists, it could be a bigger issue, like a failing thermostat or water pump.

Many people mistakenly believe overheating is only a summer problem, but cold weather can also wreak havoc on your cooling system. Frozen coolant can crack the radiator or engine block, so make sure your coolant mix includes enough antifreeze for winter. Another misconception? That coolant lasts forever. Coolant degrades over time, losing its protective properties and becoming less effective at preventing rust. Don't skip regular flushes!

Taking care of your radiator is one of the most effective ways to keep your engine running smoothly. By checking coolant levels, flushing the system regularly, and inspecting hoses, you'll avoid costly breakdowns and extend your car's lifespan. Plus, it's a straightforward job you can handle yourself without a trip to the mechanic.

RECAP

Engine maintenance can feel intimidating at first, but as we've seen in this chapter, many of the most essential tasks are surprisingly manageable. Whether it's replacing spark plugs or inspecting ignition wires, these simple actions go a long way in improving your car's performance and keeping it reliable. Taking care of these small but crucial components helps ensure smooth starts, better fuel efficiency, and fewer unexpected repairs.

We also explored how cleaning parts like the throttle body and mass airflow sensor can restore lost power and improve your engine's overall efficiency. These tasks don't require a professional touch—just the right tools and a little patience. And then there's radiator care, an often-overlooked but vital aspect of engine health. Regularly checking coolant levels and flushing the system not only prevents overheating but also saves you from costly damage to the engine block or gaskets.

Each of these steps, while simple, adds up to significant benefits. You're not just maintaining your car; you're investing in its longevity and performance. More importantly, you're gaining confidence in your ability to handle your car's care—one maintenance task at a time.

Now that you've mastered the basics of engine upkeep, it's time to turn your attention to another essential system: the electrical system. From the battery that powers it all to the intricate wiring behind your headlights and dashboard, the next chapter will guide you through understanding and managing the electrical components that keep your car running smoothly.

CHAPTER 8

THE ELECTRICAL SYSTEM UNLOCKED

REPLACING A FUSE IS AS SIMPLE AS SWAPPING A LIGHT BULB—DON'T LET IT STUMP YOU

Have you ever wondered what keeps your car's lights shining bright, your windows rolling up and down, or your engine roaring to life at the turn of a key? It's all thanks to your car's electrical system—a network of components quietly working behind the scenes to power everything from your headlights to your radio. While it might seem like a maze of wires and circuits, understanding and maintaining this system is easier than you'd think.

In this chapter, we'll demystify the essentials of your car's electrical system, focusing on three key components: the battery, fuses, and relays. You'll learn how to spot the signs of a weak or failing battery and replace it in a few straightforward steps. We'll also cover how to troubleshoot and swap out blown fuses and faulty relays, which are often the culprits behind non-working lights, radios, or power windows.

But electrical care isn't just about fixing the obvious. We'll also show you how to diagnose trickier issues using tools like a multimeter, helping you track down problems like dim headlights, flickering interior lights, or a car that just won't start. Once you know where to look, solving these problems becomes much less daunting.

And because visibility and safety are crucial, we'll guide you through light maintenance—like replacing burnt-out bulbs and restoring cloudy headlight lenses. Plus, you'll pick up tips for fixing wiring problems and keeping your electrical system running smoothly.

By the end of this chapter, you'll feel confident handling basic electrical repairs and troubleshooting. Let's roll up our sleeves, break it down step by step, and make sure your car is powered, functional, and ready for whatever the road throws at you!

REPLACING BATTERIES, FUSES, AND RELAYS

Your car's electrical system might feel like an intimidating mess of wires and circuits, but when you break it down, it relies on just a few key components: batteries, fuses, and relays. These parts are essential for everything from starting your engine to powering your headlights and radio. The best part? Maintaining and replacing them is far easier than most people think.

Understanding the Role of the Battery

At the heart of your car's electrical system is the battery. It provides the surge of energy needed to start your engine and keeps essential systems running when the car isn't moving. However, batteries aren't immortal—they wear out over time, especially in extreme temperatures or with infrequent use. Spotting a failing battery early can save you from being stranded with a dead car. So, how can you tell if your battery is failing? Look out for these signs:

- **Slow engine crank:** If your engine struggles to start, the battery may be losing its charge.
- **Dim headlights:** If your lights dim at idle but brighten when driving, the battery might be weakening.
- **Dashboard warning lights:** The battery or alternator warning light is your car's way of asking for attention.
- **Corrosion on terminals:** White or green buildup around the battery terminals disrupts the electrical flow.
- **Frequent jump-starts:** Needing a jump more than once is a red flag that your battery is nearing the end.

How to Replace a Car Battery

The good news? Swapping out a battery is one of the simplest car maintenance tasks you can do yourself. Here's how:

1. **Turn Off the Engine:** Ensure your car is off, in park, and the parking brake is engaged.
2. **Locate the Battery:** Most batteries are under the hood, but some cars tuck them in the trunk or under the rear seat. Check your owner's manual.
3. **Disconnect the Terminals:** Always disconnect the negative terminal (-) first to avoid short circuits, then the positive (+).
4. **Remove the Battery:** Carefully lift it out—it's heavier than it looks!
5. **Install the New Battery:** Place the new battery in the tray and reconnect the positive terminal first, then the negative. Tighten the connections.

6. **Secure and Test:** Make sure the battery is snug, then start the car to confirm everything works.

Pro Tip: Take your old battery to an auto parts store—they'll recycle it for free, and many offer discounts when you return it.

Troubleshooting and Replacing Fuses

Fuses are like tiny safety guards for your car's electrical system. When there's an electrical overload, a fuse "blows" to protect the circuit and prevent damage. It's a simple but effective safeguard. If a specific system, like your headlights, radio, or power windows, suddenly stops working, a blown fuse is often the culprit. Here's how to replace a fuse in just a few easy steps:

1. **Locate the Fuse Box:** Most cars have two fuse boxes—one under the hood and another inside the cabin, usually near the driver's footwell or glove box. Check your owner's manual for their exact locations.
2. **Identify the Blown Fuse:** Use the diagram on the fuse box cover or your manual to find the correct fuse. A blown fuse typically looks burnt or has a broken filament inside.
3. **Replace the Fuse:** Grab a fuse with the same amperage rating (e.g., 10A, 15A) and pop it in. Using a higher-rated fuse might seem tempting, but it can damage your electrical system or even cause a fire.
4. **Test the Circuit:** Turn on the affected system—like your lights or radio—to confirm the fuse replacement solved the issue.

Pro Tip: Keep a small fuse kit in your glove box. It's inexpensive, takes up no space, and can get you back on the road quickly without the need for a mechanic.

Troubleshooting and Replacing Relays

Relays act like switches that control high-current systems, such as the cooling fan, headlights, fuel pump, and horn. When a relay fails, these critical systems stop working, and the issue can easily be mistaken for a more complex—and expensive—problem. The good news? Relays are relatively easy to identify and replace. Here's how to replace a faulty relay in just a few steps:

1. **Locate the Relay Box:** It's usually near the fuse box under the hood. Your owner's manual will pinpoint the exact location.
2. **Identify the Faulty Relay:** A quick trick is to swap it with a working relay of the same type. For example, you can use the horn relay to test the cooling fan circuit.
3. **Replace the Relay:** Once you confirm it's the issue, simply pull out the faulty relay and plug in a new one with the correct specifications.

So, how do you know a relay is the culprit? Watch for these telltale signs:

- The system it controls stops working—like headlights that won't turn on or a silent fuel pump.
- You hear a faint clicking sound from the relay box when you try to activate the system.
- The system works intermittently, cutting in and out at random.

Pro Tip: Take a minute to label your relays when you're under the hood. A little prep now will save you time and frustration the next time an issue pops up.

With relays checked and sorted, let's move on to some common electrical troubleshooting tips that can help you identify problems with minimal guesswork.

DIAGNOSING ELECTRICAL ISSUES WITH BASIC TOOLS

Electrical issues can be frustrating, but having the right tools and a little know-how can make diagnosing them surprisingly simple. Let's go over the essential tools you'll need and how they help you identify common problems.

Tools You'll Need

You don't need a toolbox full of fancy gadgets—just a few key tools:

- **Multimeter**: This is your go-to tool for checking voltage, current, and resistance. It's invaluable for testing battery health and alternator output and identifying wiring problems.
- **Test Light**: This simple tool helps you check for power in circuits, quickly pointing out if there's a fault, like a blown fuse or broken connection.
- **Wire Strippers and Crimpers**: If you discover damaged wires, these tools let you strip insulation and make secure connections.
- **Electrical Tape and Heat Shrink Tubing**: Use these to insulate and protect wires after repairs, ensuring no exposed connections cause shorts.

Now that you've got the right tools in hand, it's time to roll up your sleeves and tackle those electrical problems. I know—car electrical issues can feel a bit overwhelming at first. But here's the thing: most of the time, the root cause is something simple, like a weak battery, a blown fuse, or a stubborn relay. Once you know where to look, troubleshooting becomes much easier, and you'll save yourself the headache (and cost) of running to a mechanic.

Let's take it step by step and break down some of the most common electrical problems, along with straightforward ways to troubleshoot them like a pro.

Common Electrical Problems and Troubleshooting Tips

Electrical issues can feel overwhelming, but most are easier to resolve than they appear. The trick is knowing where to look first. Below are some of the most common electrical problems and actionable steps to troubleshoot them.

- **Car Won't Start:** A dead battery is often the prime suspect when your car refuses to start. Use a multimeter to test the battery voltage—it should read about 12.6 volts when fully charged. If it's below 12.0 volts, the battery is weak and likely needs charging or replacing. If the battery checks out, your next steps are to inspect the starter relay and ignition switch, which can also cause starting issues.
- **Lights Flickering or Dim:** Dim or flickering lights can signal loose or corroded battery connections. Start by cleaning the terminals and tightening the connections. If the lights are still misbehaving, it's time to check the alternator. A failing alternator can't keep up with your car's electrical demands, leaving your headlights struggling to stay bright.
- **Accessories Not Working:** When electrical accessories like the radio, power windows, or interior lights stop working, blown fuses or faulty relays are the most common culprits. Check the relevant fuse and relay first. If you find a blown fuse, replace it with one of the same amperage. Never use a higher-rated fuse—it can lead to electrical damage or even fires.
- **Battery Draining Too Quickly:** If your battery seems to die overnight or shortly after turning off the car, you might have a parasitic draw—an electrical component draining power while the car is off. To find the culprit, use a multimeter to measure the current draw. Start pulling fuses one by one until the draw drops, which will point you to the problematic circuit. Dome lights, stereos, and alarm systems are common troublemakers.
- **Power Windows Stuck:** If a power window refuses to move, start with the basics. Check the fuse box for any blown fuses related to the windows. If the fuse is fine, test the window switch by swapping it with another working switch (if possible). If the window still won't budge, the motor inside the door panel may need replacing.
- **Horn Not Working:** A silent horn can be frustrating but is usually easy to diagnose. Start with the fuse and relay—swap out the horn relay with a similar one, like the cooling fan relay, to test it. If the horn still doesn't work, check the wiring and use a multimeter to ensure the horn is receiving power.
- **Check Engine Light:** When the check engine light comes on, it's easy to assume the worst. However, it often signals minor electrical issues like a dirty MAF sensor, failing oxygen sensor, or misfires. Use an OBD-II scanner to retrieve the error code, which will point you to the specific issue. From there, you can inspect, clean, or replace the affected part as needed.

- **Blown Headlights or Taillights:** If your headlights or taillights frequently burn out, there might be a wiring issue or excessive voltage. Start by inspecting the bulb sockets for corrosion or damage. Next, use a multimeter to check the voltage—anything above 12 volts can cause premature bulb failure. If everything looks fine, opt for high-quality bulbs designed for longer life.
- **Dashboard Warning Lights Stay On:** When warning lights like the battery, ABS, or airbag lights refuse to turn off, the issue often lies in a faulty sensor or control module. Begin by testing the battery and alternator to rule out voltage issues. If those are fine, use an OBD-II scanner to pull error codes and pinpoint the problem.
- **Blower Fan Not Working:** If your heater or AC blower fan stops working, start by checking the fuse and relay for the blower motor. If those are intact, the blower motor resistor could be the problem. This component often fails and prevents the fan from operating at certain speeds.

Most electrical problems can be traced back to the basics: a weak battery, loose connections, blown fuses, or faulty relays. Start small and work your way up. With a multimeter, a basic fuse kit, and a little patience, you can troubleshoot and resolve most issues yourself.

FIXING HEADLIGHTS, TAILLIGHTS, AND TURN SIGNALS

Your car's lights might not get much thought until they stop working, but they're essential for both your safety and visibility on the road. Whether it's a burned-out bulb, foggy headlight lenses, or turn signals acting up, most light-related fixes are quick, inexpensive, and totally manageable at home.

Replacing Bulbs

Changing a light bulb in your car is one of the simplest DIY tasks you can tackle. You don't need any fancy tools, just a little patience and attention to detail. Here's how to get it done:

1. **Access the Bulb Housing**: Open the hood or trunk, depending on the light you're replacing. Some cars require you to remove screws or panels to access the bulb, so check your owner's manual for guidance.
2. **Remove the Old Bulb**: Gently twist the bulb counterclockwise and pull it out. Be careful not to force it—bulb sockets can be fragile.
3. **Install the New Bulb**: Handle the new bulb by its base, not the glass. Oils from your fingers can heat unevenly on the bulb and shorten its lifespan. Insert the bulb, twist it clockwise, and make sure it's secure.
4. **Test the Light**: Before reassembling the housing, turn on your lights to confirm the new bulb works.

Pro Tip: Always keep spare bulbs in your car for emergencies. They're inexpensive, and you'll thank yourself the next time a light goes out unexpectedly.

Restoring Foggy Headlight Lenses

Headlight lenses naturally become cloudy over time due to UV exposure and wear, which can dim your lights and reduce visibility. Thankfully, you don't need to replace the entire assembly—cleaning them is surprisingly easy.

1. **Use a Restoration Kit**: Headlight restoration kits are available at auto stores and typically include polishing compound and sanding tools. Follow the instructions carefully for the best results.
2. **DIY Cleaning**: In a pinch, mix baking soda and water to form a paste. Rub it onto the lens with a microfiber cloth, rinse, and dry. Repeat if necessary.
3. **Polish and Protect**: Finish with a plastic polish or sealant to protect the lens and keep it clear longer.

Restoring foggy headlights can make a huge difference in visibility, especially for night driving.

Fixing Turn Signals

Faulty turn signals can be more than a nuisance—they're a safety hazard. When they stop working, troubleshooting is fairly straightforward.

1. **Check the Bulb**: If your turn signal blinks rapidly or doesn't light up at all, the bulb is likely burned out. Replace it following the steps for bulb replacement above.
2. **Inspect the Flasher Relay**: If the bulb is fine but the signal blinks erratically (or not at all), the flasher relay might be the culprit. Relays are inexpensive and easy to swap out—just locate the relay box, find the faulty relay, and plug in a new one.
3. **Verify the Switch**: If replacing the bulb and relay doesn't solve the problem, the turn signal switch itself might be the issue. At this point, professional help may be needed.

Addressing Wiring Issues

Sometimes, a new bulb or relay doesn't solve the issue, which points to a wiring problem. Damaged wires can interrupt power to your lights and cause frustrating malfunctions. Here's how to troubleshoot and repair wiring issues:

- **Inspect for Visible Damage**: Look for frayed, cracked, or corroded wires around the bulb housing and connectors.
- **Repair Damaged Sections**: Cut out damaged wire sections, splice in new wire, and secure the repair with heat shrink tubing to protect it from moisture and wear.
- **Test the Circuit**: Use a multimeter to check for continuity. If there's no power, trace the wiring back to the fuse or relay to pinpoint the fault.

While these fixes are simple, safety is always the priority when working on electrical systems. Here are a few safety tips you need to keep in mind when working on your car's electrical component

- **Disconnect the Battery**: Always disconnect the negative terminal before starting work to prevent shorts or sparks.
- **Wear Protective Gear**: Gloves and safety goggles can protect against unexpected shocks or debris.
- **Avoid Wet Conditions**: Electricity and moisture don't mix, so work on a dry day.
- **Follow the Manual**: Your owner's manual is your best resource for car-specific instructions.

Fixing your car's headlights, taillights, and turn signals doesn't require a trip to the shop. With a little know-how, you can save money and ensure your lights are always shining bright. Whether you're replacing a bulb, cleaning foggy lenses, or fixing wiring issues, these small tasks give you the confidence to take charge of your car's care. Plus, they're quick wins that can keep you—and everyone else on the road—safe and visible.

RECAP

At first glance, your car's electrical system might seem like a confusing tangle of wires and circuits, but as we've uncovered in this chapter, it's far less intimidating when broken down into its core components. You've learned the essentials, from understanding the battery's vital role in powering your car to replacing fuses and relays when something goes wrong. These simple tasks give you the ability to troubleshoot and resolve common electrical issues that can disrupt your car's performance.

We also explored the value of diagnostic tools like a multimeter. With it, you can identify the root cause of problems such as dim headlights, a draining battery, or malfunctioning accessories. And then there are the small but impactful fixes—like replacing burnt-out bulbs or restoring foggy headlight lenses—that not only improve visibility but also boost your safety on the road. These quick wins save you time, money, and the hassle of unnecessary trips to the mechanic.

Each step you've learned not only simplifies car care but also builds your confidence in managing your vehicle's electrical system. Knowing how to diagnose and fix these common issues empowers you to keep your car running smoothly, avoid costly repairs, and stay prepared for unexpected challenges.

Now that you're equipped with the tools and know-how to maintain your car's electrical system, it's time to shift focus to another critical part of your vehicle: the transmission and drivetrain. These systems are the heart of what keeps your car moving, and in the next chapter, we'll uncover the secrets to keeping them in top shape. Let's keep building your expertise and ensure every part of your car is ready for the road ahead!

CHAPTER 9

TRANSMISSION AND DRIVETRAIN ESSENTIALS

AUTOMATIC OR MANUAL, YOUR TRANSMISSION DESERVES REGULAR TLC

Did you know that a small issue in your drivetrain can lead to uneven tire wear, sluggish acceleration, or even higher fuel consumption? These systems might not grab your attention like your engine or brakes, but your drivetrain and transmission are the backbone of every drive. They quietly work together to deliver the power your engine generates directly to your wheels, keeping your car running smoothly and efficiently.

In this chapter, we'll break down the key components of the drivetrain—like the driveshaft, differential, CV joints, and axles—and explain how they work in harmony to keep your car moving. You'll also learn how to recognize early warning signs of drivetrain trouble, such as vibrations, clunking noises, or fluid leaks. Addressing these signs early can prevent small problems from becoming costly repairs.

But it's not just about the drivetrain. Your car's transmission plays a huge role in managing how that power gets to the wheels. Whether you're driving an automatic or manual transmission, we'll cover how they work, why transmission fluid is critical, and how to maintain it. Plus, we'll highlight the telltale signs of transmission trouble so you can take action before it becomes a major headache.

Finally, we'll focus on inspecting and maintaining CV joints and axles. These often-overlooked parts are critical for smooth handling and transferring power to your wheels. With some basic knowledge and care, you'll be able to keep these components in top condition, saving you money and avoiding unnecessary breakdowns.

By the end of this chapter, you'll have a clear understanding of how these systems work, how to keep them running smoothly, and how to spot problems early. Let's dive into the mechanics of what keeps your car moving forward!

UNDERSTANDING THE DRIVETRAIN SYSTEM

The drivetrain is one of the most vital systems in your car. It works alongside the engine and transmission to transfer power to the wheels, allowing your vehicle to move. Think of it as the bridge that connects the power generated by the engine to the tires on the road. A properly functioning drivetrain ensures smooth and efficient power delivery, whether you're accelerating, cruising, or navigating tricky terrain.

What Makes Up the Drivetrain?

To understand how the drivetrain works, it's helpful to break it down into its key components. Depending on whether your car is front-wheel, rear-wheel, all-wheel, or four-wheel drive, the setup can vary slightly. Here are the main parts you should know:

1. **Transmission**: Sends power from the engine to the driveshaft.
2. **Driveshaft**: A rotating shaft that transfers power from the transmission to the differential.
3. **Differential**: Allows the wheels to rotate at different speeds, particularly during turns.
4. **CV Joints (Constant Velocity Joints)**: Found in front-wheel-drive cars, they connect the transmission to the wheels and allow smooth rotation, even at sharp angles.
5. **Axles**: Deliver power from the differential to the wheels.
6. **Transfer Case** *(in AWD/4WD)*: Splits power between the front and rear wheels.

Each of these components plays a unique role, but they work together as a team. When one part begins to fail, the entire drivetrain can suffer—and that's when you'll start noticing problems.

Signs of Drivetrain Issues

So, how do you know when something's wrong with your drivetrain? The symptoms often reveal themselves through unusual sounds or changes in your car's performance. If you pay close attention, you can spot these signs early:

- **Strange Noises:** Unusual sounds are often the first clue that something isn't right.
  - **Clunking Sounds**: Worn-out universal joints (U-joints) in the driveshaft can create loud clunks.
  - **Grinding or Whining**: A grinding noise during acceleration may signal trouble with the differential or a worn CV joint, while whining often points to low transmission or differential fluid.

  - **Humming or Groaning**: A humming sound that grows louder as you accelerate could mean an issue with the wheel bearings or axles.
- **Vibrations While Driving:** If you feel a vibration through the steering wheel, floor, or seat, it's often tied to the driveshaft or CV joints. Worn U-joints or an unbalanced driveshaft can create shaking at certain speeds, which gets worse over time.
- **Difficulty Shifting Gears:** Hesitating, jerking, or struggling to shift gears smoothly? This may point to transmission trouble. Low transmission fluid, worn components, or a faulty transfer case can make gear shifts rough or delayed.
- **Unusual Tire Wear:** Keep an eye on your tires. Uneven or excessive tire wear can signal drivetrain misalignment, axle problems, or a failing differential. If one wheel isn't receiving power correctly, it might drag, causing abnormal wear patterns.
- **Unresponsive Acceleration:** If pressing the gas pedal gives you delayed or weak acceleration, the drivetrain might not be transferring power efficiently. This issue often stems from a slipping transmission, damaged axles, or a failing differential.
- **Fluid Leaks:** Leaking fluid is never a good sign. Dark or red fluid under the car usually means trouble with the transmission, differential, or transfer case. These components rely on proper lubrication, so address leaks quickly to prevent costly internal damage.

If you notice any of these symptoms, it's a good idea to address them sooner rather than later. Drivetrain issues tend to worsen over time, turning small problems into expensive repairs. Start by inspecting the driveshaft, CV joints, and axles for visible damage or leaks. Check your fluid levels regularly and top them off as needed.

UNDERSTANDING YOUR CAR'S TRANSMISSION: AUTOMATIC VS. MANUAL

Your car's transmission might seem like a complex black box, but its role is straightforward—it ensures the engine's power gets to your wheels efficiently. Think of it as the middleman, managing how power is distributed so your car can accelerate, climb hills, or cruise at highway speeds. Whether you drive an automatic or manual transmission, understanding how it works and recognizing early signs of trouble can save you a lot of headaches and repair bills.

That said, what exactly does a transmission do? In essence, the transmission adjusts the power coming from your engine to suit your driving needs. It uses different gear ratios to deliver torque (rotational power) for acceleration or fuel efficiency. Lower gears give you the force to get moving, while higher gears allow for smoother cruising with less engine effort.

Automatic Transmission

If you drive a car with an automatic transmission, you've got it easy. The system does all the gear shifting for you using hydraulics, electronics, and sensors. It automatically detects your driving speed, load, and other conditions, then shifts gears seamlessly.

Automatic transmissions offer several advantages that make driving easier and more convenient, especially for beginners or those in heavy traffic.

- First, **simplicity is key**. You don't need to worry about shifting gears manually, which is a lifesaver in stop-and-go traffic or during long commutes. The transmission handles it all for you.
- Next is **smooth operation**. Modern automatics are engineered for seamless, effortless gear changes, providing a more comfortable driving experience.
- Finally, **convenience** makes automatics even more appealing. They dominate the market, meaning vehicles equipped with automatic transmissions—and their replacement parts—are widely available and easy to find.

To keep an automatic transmission running smoothly, regular maintenance is essential. Here's how to go about it:

1. **Check and Change Transmission Fluid**: Transmission fluid keeps everything lubricated and cool. Low or dirty fluid can cause slipping gears, overheating, or even failure. Check it regularly (using the dipstick if your car has one) and change it as recommended in your manual—often every 30,000 to 60,000 miles.
2. **Watch for Signs of Trouble**: Slipping gears, delays when shifting, or strange whining noises are all red flags. Don't ignore them.
3. **Inspect for Leaks**: Look for red or brown fluid under the car, especially near the transmission pan. A leaky gasket or seal can cause major issues if left untreated.

Manual Transmission

For those who love having more control over their driving experience, manual transmissions (or "stick shifts") are the way to go. Instead of relying on electronics, you shift gears yourself using a clutch pedal and gear shifter.

Manual transmissions come with several benefits that appeal to drivers who value control and efficiency.

- **Driver control** is one of the biggest advantages. You decide when to shift gears, which is particularly useful on hilly terrain or tricky roads where precision matters. This level of control makes manual driving a favorite among enthusiasts.

- Another key benefit is **lower repair costs**. Manuals are simpler in design compared to automatics, which usually means fewer components to fail and more affordable repairs.
- Lastly, there's **fuel efficiency**. Manuals can deliver better fuel economy, especially when compared to older automatic transmissions, saving you money over time.

Like automatics, manual transmissions require regular care. Here's what you need to do:

1. **Clutch Wear**: The clutch is critical in a manual transmission. Over time, it will wear out and need replacing. If you notice slipping, difficulty shifting, or grinding, it's time to check the clutch.
2. **Monitor Fluid Levels**: Manuals also rely on fluid for lubrication. Low fluid can lead to stiff gear shifts and damage.
3. **Pay Attention to Noise**: Grinding or whining noises during shifting usually indicate issues with the clutch or gearbox. Address them early.

The decision between an automatic and a manual transmission often comes down to personal preference and driving conditions. Manuals are perfect for drivers who enjoy full control and engagement, while automatics are ideal for those who value convenience, especially in heavy traffic.

Regardless of the type, one thing remains the same: regular maintenance is critical. A neglected transmission—whether automatic or manual—can turn into an expensive problem. By understanding how your transmission works and addressing issues early, you can keep your car running smoothly for years to come.

TRANSMISSION FLUID MAINTENANCE MADE SIMPLE

Transmission fluid is vital for the health of your transmission. It's the lifeblood of this system, ensuring lubrication, cooling, and smooth operation. Yet, many drivers neglect this essential fluid, often because they aren't sure when or how to maintain it. Here's everything you need to know.

To understand its importance, it helps to know what transmission fluid actually does. In automatic transmissions, it serves as both a lubricant and a hydraulic medium, making gear shifts smooth and efficient. For manual transmissions, it keeps moving parts lubricated and reduces friction. In both cases, it also plays a crucial role in cooling the system and preventing premature wear.

However, transmission fluid doesn't last forever. Over time, heat and contamination cause it to degrade, making it less effective at protecting your transmission. When that happens, issues like slipping gears or overheating can arise—and those are problems you want to avoid.

Choosing the Right Transmission Fluid

Not all transmission fluids are created equal. Using the wrong type can harm your transmission, so it's essential to stick to what your vehicle's manufacturer recommends. The two main types are:

- **Automatic Transmission Fluid (ATF):** Designed for automatic transmissions and some newer manuals.
- **Manual Transmission Fluid:** Thicker and heavier, formulated specifically for traditional manual gearboxes.

Check your owner's manual to confirm the correct fluid for your car. It's a small detail that makes a big difference.

When to Check and Replace Transmission Fluid

Most manufacturers suggest checking your transmission fluid every 30,000 to 60,000 miles, but your driving habits can impact this timeline. If you frequently drive in hot weather, haul heavy loads or tow trailers, you'll need to check and replace the fluid more often.

How to Check Transmission Fluid:

1. Park the car on a level surface and let the engine warm up.
2. Depending on your vehicle, check whether the transmission fluid should be inspected while the car is idling or when the engine is off—this varies by manufacturer. Consult your owner's manual for the correct procedure.
3. Locate the transmission dipstick, which is usually near the engine bay.
4. Pull out the dipstick, wipe it clean, then reinsert it fully.
5. Pull it out again to check the fluid's level and color.
   - **Healthy fluid:** Bright red or pink with a neutral smell.
   - **Bad fluid:** Dark, dirty, or burnt-smelling.

If your fluid looks or smells off, it's time for a replacement.

How to Replace Transmission Fluid

Replacing transmission fluid might sound intimidating, but it's a task you can handle with the right tools and precautions. Here's a quick guide:

1. Safely lift your car with a jack and secure it with jack stands.
2. Place a container beneath the transmission drain plug and remove the plug.
3. Let the old fluid drain completely.
4. Replace the drain plug and refill the system with the correct transmission fluid, using a funnel to avoid spills.
5. Recheck the fluid level after refilling and top off if necessary.

Signs Your Transmission Fluid Needs Attention

Ignoring transmission fluid can lead to noticeable problems. Watch for these warning signs:

- **Slipping gears:** It feels like your car is shifting gears on its own.
- **Rough or delayed shifts:** Lubrication issues can cause jerky or hesitant gear changes.
- **Leaking fluid:** Look for reddish fluid under your car. Leaks need immediate attention.
- **Burning smell:** Overheating fluid produces a distinct burnt odor.

INSPECTING AND MAINTAINING CV JOINTS AND AXLES

Your car's transmission might steal the spotlight, but CV joints and axles play an equally critical role in keeping you moving. These often-overlooked components ensure power gets delivered smoothly to your wheels, allowing them to turn and handle bumps, dips, and uneven surfaces.

What Are CV Joints and Axles?

Let's start with the basics.

- **CV Joints (Constant Velocity Joints):** These flexible couplings connect the axles to your wheels. They're designed to deliver consistent power, even when the wheels are turning or moving up and down with the suspension. Whether you're making a tight turn or navigating rough roads, CV joints ensure your ride stays smooth.
- **Axles:** Axles are the solid shafts that transfer power from the transmission to your wheels. They work hand-in-hand with CV joints to manage the stress of acceleration, turning, and uneven terrain.

Together, these components keep your car responsive and adaptable—so when they start to fail, you'll notice it pretty quickly.

Common Issues and Warning Signs

CV joints and axles don't just quit overnight. Instead, they give off warning signs that something's wrong:

- **Clicking or popping sounds when turning:** This is one of the biggest red flags of a worn CV joint.
- **Vibrations while driving:** A damaged or unbalanced axle can cause noticeable shaking, especially at higher speeds.
- **Grease leaks near the wheels:** A torn CV boot can let grease escape, leaving the joint unprotected and prone to failure.

The sooner you catch these symptoms, the easier—and cheaper—the fix will be.

Inspecting CV Joints and Axles

Routine inspections can help you spot trouble before it escalates. Here's how to check your CV joints and axles:

1. **Look for grease leaks:** Inspect the rubber CV boots around the joints. If you see cracks, tears, or grease splatter, the boot is likely compromised.
2. **Listen for noises:** Drive slowly in a parking lot or safe space and make sharp turns. Clicking or popping sounds are clear signs of CV joint wear.
3. **Feel for vibrations:** Pay attention to how the car feels, particularly at higher speeds. Unusual vibrations can point to axle damage or misalignment.

Replacing CV Joints and Axles

If you find significant damage—like torn boots or worn joints—replacement is usually the best solution. While this job requires a bit of mechanical know-how, it's doable with the right tools and guidance. Here's how to replace a CV joint or axle in a few short steps:

1. Lift and secure the vehicle safely with a jack and jack stands.
2. Remove the wheel to access the CV joint or axle.
3. Disconnect suspension components as necessary to free up the axle.
4. Extract the damaged CV joint or axle carefully, taking care not to damage surrounding parts.
5. Install the new component, ensuring all connections are tight and secure.
6. Reassemble everything and test the car. Drive it around slowly to check for noises or vibrations.

Taking a little time for preventative care can go a long way in keeping your CV joints and axles in great shape. It's all about catching small problems early and forming a few good habits to avoid expensive repairs later.

First, keep an eye on your CV boots. If you spot cracks or tears, don't brush it off—fix it as soon as possible. Those boots act as shields, keeping dirt and moisture away from the joint. A torn boot might not seem like a big deal at first, but once grime sneaks in, it can wear out the joint faster than you think, turning a quick fix into a costly replacement.

Second, take it easy on the gas pedal and steering wheel. Hard accelerations or sharp, aggressive turns put unnecessary stress on your CV joints. This is especially true if you drive on rough roads or regularly make tight maneuvers. Gentle driving helps everything last longer.

Finally, don't neglect your suspension. When shocks, struts, or other suspension components wear out, they can throw your axles out of alignment, causing uneven wear and extra strain on your CV joints. Keeping the suspension in good condition ensures everything works together smoothly and helps you avoid premature damage.

A little attention now means fewer headaches (and bills) down the road. These simple steps can keep your car running smoothly and save you from surprises later. Of course, if this feels out of your comfort zone, don't hesitate to consult a professional—but understanding the process will still help you make informed decisions.

RECAP

Your car's transmission and drivetrain might seem complicated at first, but as we've explored in this chapter, their roles are pretty straightforward. They work together to deliver power from the engine to your wheels, ensuring smooth and efficient movement. By understanding how key components like the driveshaft, CV joints, and axles function, you're now better equipped to spot problems early and address them before they escalate.

We also dove into the importance of transmission fluid maintenance. Keeping your transmission fluid clean and at the right levels is essential for preventing slipping gears, overheating, and other common issues. You've learned how to check and replace it, as well as what signs—like strange noises or leaks—might indicate it's time for action.

And let's not forget the vital role of CV joints and axles in your car's performance. By inspecting for signs of wear, such as grease leaks or vibrations, you can keep these components in great shape. Addressing small issues early can save you from costly repairs and ensure your car remains responsive and safe to drive.

These systems may not demand attention every day, but regular care can go a long way in maintaining your car's reliability and performance. With the insights you've gained, you're now better prepared to handle routine maintenance and troubleshoot potential issues with confidence.

Next, we'll shift gears and uncover some of the best-kept secrets in car care. The upcoming chapter will share expert tips and tricks that mechanics often keep to themselves—insider knowledge that will make you a savvier, more confident car owner. Ready? Read on.

CHAPTER 10

PRO TIPS FROM AN EXPERT MECHANIC

NOT EVERY RECOMMENDATION IS NECESSARY—LEARN TO SEPARATE THE MUST-DOS FROM THE UPSELLS

Have you ever left a repair shop with that nagging feeling you might have paid for something you didn't really need? If so, you're not alone. While many mechanics are honest and hardworking, there are some common industry practices—like upselling or charging for unnecessary services—that can leave your wallet lighter than it needs to be. The good news? A little insider knowledge can go a long way in helping you avoid these pitfalls.

In this chapter, we'll shed light on the strategies some repair shops use and show you how to protect yourself from overpaying. You'll learn about diagnostic fees and why you might not need to pay them, plus how to spot and avoid unnecessary upsells like premature filter replacements or pricey "fuel system cleanings." We'll also debunk the myth of "dealer-only parts" and show you where to find high-quality components for less.

But saving money isn't just about dodging unnecessary costs—it's also about taking control. That's why we'll share practical tips for negotiating with mechanics, like how to get multiple estimates and ask the right questions to ensure you're paying a fair price. And if you've been hesitant to try DIY repairs, we'll highlight simple tasks you can start with to build confidence and save hundreds in the process.

This isn't about being suspicious of every repair shop—it's about being informed and prepared. By the end of this chapter, you'll have the tools and knowledge to make smarter decisions about your car's care, whether you're at the shop or working on it yourself. Let's dive in and discover the secrets that will help you keep your car running smoothly while saving money!

SECRETS MECHANICS DON'T WANT YOU TO KNOW

Most mechanics are hardworking professionals who genuinely aim to help. But let's be honest: like any industry, the auto repair world has its share of tricks that can leave you overpaying for services you may not need. Knowing these secrets can help you protect your wallet and make smarter decisions when your car needs attention.

The Truth About Diagnostic Fees

Diagnostic fees are one of the most common ways repair shops boost their bottom line. These fees—ranging from $50 to $150 or more—are charged to identify issues with your car, such as when the check engine light comes on. While diagnostics are often necessary, some shops take advantage by charging you for something that's simple and quick.

Here's the truth: many auto parts stores, like AutoZone or Advance Auto Parts, will scan your car for free. All they do is connect a diagnostic tool to read error codes, which takes only a few minutes. Better yet, you can invest in an OBD-II scanner for around $30–$50 and check the codes yourself at home. Before agreeing to a paid diagnostic test, ask if there's a free option available—it could save you a bundle.

The "Upsell" Playbook

We've all been there: you go in for a quick oil change and suddenly get handed a list of "urgent" repairs worth hundreds of dollars. Upselling is a common practice where shops recommend services that aren't truly needed or are premature. For example, they might suggest replacing an air filter that still has plenty of life left or selling you on a "fuel system cleaning" that your car doesn't actually require. To avoid falling for this, always ask questions:

- *Why is this service necessary?*
- *What symptoms indicate the problem?*
- *Can I see the worn part?*

Also, consult your owner's manual. It includes a detailed maintenance schedule tailored to your car, so you'll know when services like filter replacements, brake checks, or fluid changes are genuinely needed.

The Myth of "Dealer-Only Parts"

Another tactic some repair shops use is claiming that certain parts can only be purchased from the dealership. This is often their way of justifying steep prices. In reality, "dealer-only" parts are rarely exclusive. Many independent suppliers and online retailers—like RockAuto, Amazon, or AutoZone—offer the same parts at significantly lower prices.

You can also explore aftermarket parts, which often meet or even exceed OEM (Original Equipment Manufacturer) standards. For budget-friendly alternatives, don't forget salvage yards or websites like Parts Geek, where refurbished parts can save you a fortune. Always take a moment to research and compare prices before accepting the shop's claim or quote.

Labor Time Estimates Aren't Always Accurate

Most shops calculate labor fees using standardized labor guides, which estimate how long a repair "should" take. This system is meant to ensure fairness and consistency. However, experienced mechanics often work faster than the guide suggests—but you still get billed for the full estimated time.

For example, if replacing a water pump is quoted as a three-hour job, you'll be charged for three hours, even if the mechanic finishes in two. This isn't always unfair—it's just how the industry works—but it can leave you overpaying.

To protect yourself, compare labor costs at different shops, especially if one estimate seems unusually high. Don't be afraid to ask for clarification or get a second opinion. Knowledge is power, and a little research can save you a lot of money.

Understanding these secrets doesn't mean you need to distrust every mechanic—it's about being informed and confident. By questioning diagnostic fees, staying alert to upsells, researching parts, and double-checking labor estimates, you can navigate car repairs without feeling overwhelmed. Remember, it's your car and your money. The more you know, the better choices you'll make.

HOW TO AVOID OVERPAYING FOR PARTS AND SERVICES

Car repairs can feel like a financial black hole, but with a little strategy and know-how, you can save a surprising amount of money. The trick is to stay informed, ask questions, and take charge of the process. Let's explore some simple ways to keep your repair costs under control without sacrificing quality.

Know the Value of Your Parts

One of the easiest ways to avoid overpaying is to understand the cost of parts. Before approving any repairs, always ask for a detailed estimate that breaks down parts and labor costs. Once you have the numbers, spend a few minutes researching the parts online to see if the quoted price is fair.

Mechanics often mark up parts by 25–50%—sometimes even more. If your shop allows it, sourcing the parts yourself can lead to huge savings. Here are a few tips for buying parts:

- Stick with OEM (Original Equipment Manufacturer) or high-quality aftermarket options to ensure reliability.
- Check trusted online retailers like RockAuto, Amazon, or eBay Motors for discounts.
- For hard-to-find or expensive components, local salvage yards can be goldmines.

Just make sure any parts you purchase are compatible with your vehicle and meet quality standards.

Skip the Dealership (When Possible)

Dealerships often have the highest repair costs, especially for vehicles that are no longer under warranty. While they might offer specialized tools and expertise, most routine repairs don't require dealer-level service. For example:

- Simple tasks like **oil changes, brake pad replacements, and battery swaps** are easy to handle at independent shops—or even on your own—and can save you hundreds.
- Save the dealership visits for warranty-covered work or extremely complex repairs that require manufacturer-specific knowledge.

Independent mechanics can often do the same job at a fraction of the cost without compromising quality.

Timing is Everything

Believe it or not, *when* you schedule repairs can affect the price you pay. Repair shops tend to be busiest during the spring and summer when people hit the road for vacations, and warmer weather causes more wear and tear. During these peak seasons, discounts are rare, and shops are often swamped.

If possible, aim for late fall or winter when business slows down. Shops are more likely to offer promotions or negotiate better rates during their off-peak times.

Preventive Maintenance Saves Money

A little preventative care today can save you thousands in future repairs. Simple, routine maintenance keeps your car running smoothly and prevents minor issues from turning into major headaches.

For instance:

- Replacing a worn **timing belt** on schedule can prevent catastrophic engine damage that costs thousands to repair.
- Keeping your **tires properly inflated** and rotated extends their life and improves fuel efficiency.

Track your maintenance with a log or a simple notebook. Knowing what's been done and what's due next will keep you ahead of the game.

Consider DIY Repairs

Don't underestimate yourself—many car repairs are easier than you think. You don't need to be a professional mechanic to tackle common tasks like replacing windshield wipers, cabin air filters, or even brake pads. Thanks to online tutorials and step-by-step videos on platforms like YouTube, learning to DIY has never been more accessible. If you're new to DIY repairs:

1. Start small with easy projects, like changing your wiper blades or air filter.
2. Invest in a basic toolkit with essentials like a socket set, pliers, and screwdrivers.
3. Build your confidence over time by gradually taking on more complex tasks.

Not only will you save money, but there's also a unique satisfaction that comes from fixing your own car. Plus, you'll develop skills that will serve you for years to come.

TIPS FOR NEGOTIATING WITH MECHANICS WHEN NECESSARY

Even if you're a DIY enthusiast, there will come a time when professional help is unavoidable. Whether it's a complex repair or a task you simply don't have the tools for, knowing how to navigate the conversation with a mechanic can make all the difference. These tips will help you get the best value for your money and ensure you leave the shop feeling confident and not taken advantage of.

Always Get Multiple Estimates

When you're facing a big repair bill, don't settle for the first quote you receive. Shop around and request estimates from at least three different repair shops. Not only does this give you a better understanding of the market rate for parts and labor, but it also puts you in a stronger position to negotiate. For example, if one shop quotes you $800 to replace a part and another offers the same job for $600, you can use that knowledge to your advantage.

Build a Relationship with Your Mechanic

Trust goes a long way in the auto repair world. If you find a mechanic who's reliable and honest, stick with them. Loyal customers often receive benefits like priority scheduling, discounts, and straight-shooting advice on which repairs are urgent and which can wait. Building this trust can be as simple as treating your mechanic with respect and showing appreciation for their work. A little courtesy can go a long way.

Use Knowledge to Your Advantage

Before you head to the shop, take a few minutes to do some research. Understand your car's symptoms, possible causes, and the general price range for repairs. If you know replacing an alternator typically costs $400–$600, you'll spot an unreasonable quote right away. Mechanics are far less likely to inflate costs when they realize you're an informed customer who knows what's fair.

Don't Be Afraid to Say No

It's easy to feel pressured when a mechanic suggests additional repairs or services you didn't ask for. But remember—it's your car and your money. If they recommend something that doesn't seem necessary, politely decline and stick to the original scope of work. Don't let them upsell you on a "fuel system cleaning" or new air filter unless you're sure it's needed.

Ask for Discounts or Promotions

You'd be surprised how many shops offer discounts—you just have to ask. Whether it's military, student, or senior citizen discounts, seasonal promotions, or a small price break for cash payments (to avoid credit card fees), these savings can add up. A simple "Do you have any discounts or specials running right now?" could knock a chunk off your bill.

Negotiate Labor Costs

Labor costs are often the biggest part of a repair bill, but they're not always set in stone. If you've sourced your own parts or have estimates from other shops showing lower labor rates, let your mechanic know. Many shops are willing to adjust their rates or meet you halfway to win your business.

Negotiating with mechanics doesn't mean being difficult—it's about being informed, respectful, and confident.

RECAP

This chapter pulled back the curtain on some of the secrets that can save you time, money, and stress when dealing with car repairs. From understanding diagnostic fees to spotting unnecessary upsells, you now have the tools to navigate the repair process with confidence. Armed with strategies like sourcing your own parts, avoiding dealership markups for routine repairs, and timing your visits to take advantage of off-peak discounts, you can take control of your car care expenses.

We also highlighted the importance of preventative maintenance and DIY repairs. Tackling small tasks—like replacing air filters, wiper blades, or brake pads—not only saves you money but also helps prevent larger, costlier problems down the road. For those repairs that do require professional help, you now know how to negotiate effectively, get multiple estimates, and ask the right questions to ensure you're getting a fair deal.

More than anything, this chapter emphasized the value of being an informed car owner. When you're equipped with the right knowledge, you're not just a passive customer—you're an active participant in your car's care. That perspective empowers you to protect both your wallet and your vehicle's longevity.

Now that you're armed with these pro tips, it's time to put them into action. In the next chapter, we'll get hands-on with 30 practical DIY repair projects you can do at home. From small fixes to slightly more advanced tasks, you'll learn how to save money and boost your confidence with every completed project.

CHAPTER 11

30 DIY REPAIR PROJECTS YOU CAN DO AT HOME

SAVE MONEY AND BUILD CONFIDENCE —ONE DIY PROJECT AT A TIME

Did you know that replacing your own brake pads can save you more than $300? Or that changing a cabin air filter takes under 10 minutes and costs a fraction of what a repair shop might charge? These simple, practical fixes are just the start of what you can accomplish with a little know-how and the right tools—no mechanic required.

In this chapter, we'll explore 30 DIY repair projects that you can tackle at home, from everyday maintenance to more involved repairs. Whether you're a complete beginner or someone eager to expand your skills, these projects are designed to be approachable and rewarding. You'll learn to handle basics like oil changes, tire rotations, and wiper blade replacements—tasks that keep your car running smoothly and safely. We'll also guide you through slightly more advanced repairs, like replacing spark plugs, cleaning your mass airflow sensor, and restoring cloudy headlights. These projects can significantly improve your car's performance and appearance without breaking the bank.

Each repair is broken down into straightforward, actionable steps, with tips to help you avoid common mistakes and troubleshoot when needed. You'll also find guidance on gathering the right tools, staying safe, and working efficiently. Along the way, you'll build confidence in your abilities, save money on professional labor costs, and gain a deeper appreciation for how your car operates.

By the end of this chapter, you'll be equipped with practical skills that empower you to take control of your car's care. So, grab your tools, clear a spot in the driveway, and let's dive into repairs you'll soon wonder why you ever paid someone else to handle!

INTRODUCTION TO DIY CAR REPAIR PROJECTS

Owning a car doesn't mean you need to run to the mechanic every time something seems off. The truth is, plenty of car maintenance tasks can be done right in your own driveway—and often for far less than what you'd pay at a shop. With a bit of time, the right tools, and clear instructions, you'll be surprised at what you can accomplish. Not only will you save money, but you'll also keep your car running like a dream.

This chapter is here to be your step-by-step guide to mastering 30 DIY car repair projects. These aren't just money-saving fixes—they're confidence boosters. Imagine changing your own oil, replacing worn-out brake pads, or finally tackling that pesky dashboard warning light. It's all doable, whether you're a total beginner or someone who's already a little hands-on.

Now, don't worry—this isn't about turning you into a professional mechanic overnight. Instead, think of these projects as essential life skills that put you in control of your car's health. And let's be honest—there's nothing more satisfying than fixing something with your own two hands and knowing you did it right.

We've set up this chapter to ease you in gently. You'll start with simple, beginner-friendly projects, like checking tire pressure or swapping out your windshield wipers. From there, you can build your confidence and move on to bigger tasks, like replacing a serpentine belt or inspecting your brakes. For each project, you'll find:

- A clear list of tools you'll need.
- An estimate of how much time it'll take.
- Safety tips to keep both you and your car in good shape.
- Troubleshooting advice to avoid common mistakes.

By the time you've worked through these projects, you'll have the knowledge and skills to handle your car with confidence. So grab your tools, roll up your sleeves, and let's get started!

TOOLS AND SAFETY TIPS FOR SUCCESSFUL REPAIRS

Before diving into any repair, preparation is half the battle. Having the right tools and following a few essential safety steps can mean the difference between a smooth, satisfying job and a frustrating one. Plus, a little planning ensures your project stays safe, efficient, and even enjoyable.

Essential Tools for DIY Car Repairs

While Chapter 2 provides an exhaustive list of tools every DIYer should have, here's a quick reminder of the basics you'll need for the projects in this chapter:

- **Socket set** (both metric and SAE sizes)
- **Wrenches** (adjustable and combination wrenches)
- **Screwdrivers** (both flathead and Phillips)
- **Jack and jack stands** (or a sturdy set of ramps)
- **Torque wrench** (for accurate bolt tightening)
- **Pliers** (regular and needle-nose)
- **Multimeter** (a must for diagnosing electrical issues)
- **Oil filter wrench**
- **Funnel and drain pan**
- **Rubber gloves and safety goggles**
- **Work light or flashlight** (you'll appreciate it in low-light spots)

These tools will handle most of the projects ahead, and investing in quality basics now will save you countless headaches later.

Safety Tips to Keep in Mind

No repair is worth putting yourself at risk, so safety always comes first. Here are a few tips to ensure you stay safe while working on your car:

1. **Work on a flat, stable surface**: Park your car on level ground and secure it with wheel chocks if needed. Stability is key to safety.
2. **Use jack stands**: Never trust a jack alone to hold your car. Always support the car with jack stands or sturdy ramps to avoid accidents.
3. **Disconnect the battery**: Before doing any electrical work, disconnect the negative battery terminal to prevent shocks or short circuits.
4. **Keep a fire extinguisher nearby**: This is especially important when working near fuel systems, the engine, or anything that generates heat. Better safe than sorry.
5. **Stay organized**: As you disassemble components, keep bolts, nuts, and small parts labeled or laid out in order. This small habit will save you a lot of frustration during reassembly.

By setting yourself up with the right tools and prioritizing safety, you'll make each repair far more manageable—and way less stressful. Now that you're ready let's get to work!

BEGINNER PROJECTS: SIMPLE TASKS TO GET STARTED

These foundational tasks are quick, easy, and great for building confidence.

Checking and Changing Engine Oil

Changing your car's engine oil is one of the simplest and most cost-effective maintenance tasks you can do at home. It's also one of the most critical. Oil keeps your engine lubricated, preventing friction between moving parts and helping dissipate heat. Over time, oil breaks down and collects dirt, which reduces its effectiveness. Regular oil changes ensure your engine runs smoothly and lasts longer.

Tools Needed:

- Socket wrench
- Oil filter wrench
- Drain pan
- Funnel
- New oil filter
- Engine oil (check your manual for type and quantity)
- Gloves and shop towels

Estimated Time: 30–45 minutes

Instructions:

- **Warm Up the Engine:** Start your car and let the engine run for about five minutes. This warms up the oil, making it thinner and easier to drain. Turn the engine off and let it cool slightly—working with piping-hot components can lead to burns.
- **Lift and Secure the Car:** Use a jack to carefully lift your car and place jack stands under the recommended lifting points. Never rely on a jack alone—always secure the car with jack stands to avoid accidents. Ensure you're working on a level surface for safety and proper drainage.
- **Drain the Old Oil:** Locate the oil pan underneath your car (it's typically a metal pan near the engine). Position your drain pan directly under the drain plug. Use a socket wrench to loosen and remove the plug. Allow the oil to drain completely—this might take a few minutes. Once done, reattach the drain plug and tighten it securely, but don't overtighten.
- **Replace the Oil Filter:** Using an oil filter wrench, unscrew the old filter. It's likely to have oil in it, so keep the drain pan nearby. Before installing the new filter, apply a thin layer of fresh oil to the rubber gasket on the filter—this creates a good seal and prevents it from sticking. Screw the new filter on hand-tight.

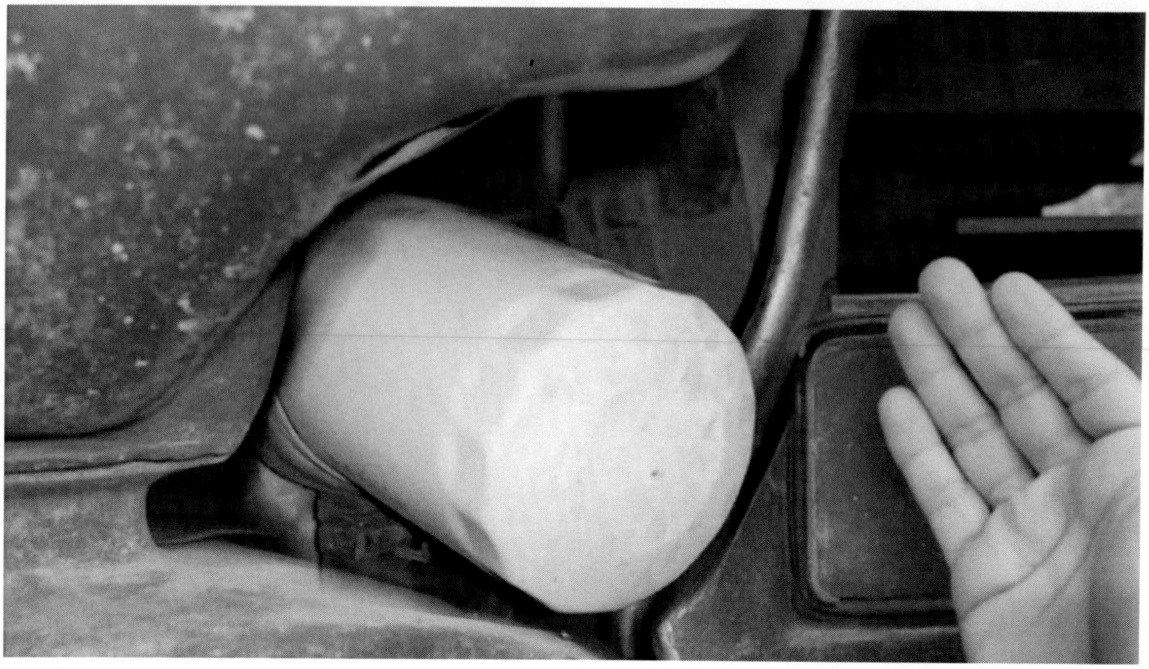

- **Add New Oil:** Pop the hood and remove the oil cap on top of the engine. Use a funnel to pour in the recommended type and amount of oil. Refer to your car's manual for precise instructions.
- **Check the Oil Level:** Once the new oil is in, start the engine for about 30 seconds. This allows the oil to circulate. Turn the engine off and wait a minute. Then, use the dipstick to check the oil level. Pull it out, wipe it clean, and reinsert it fully. Pull it out again and ensure the oil level is between the "min" and "max" marks. If it's low, add more oil as needed.

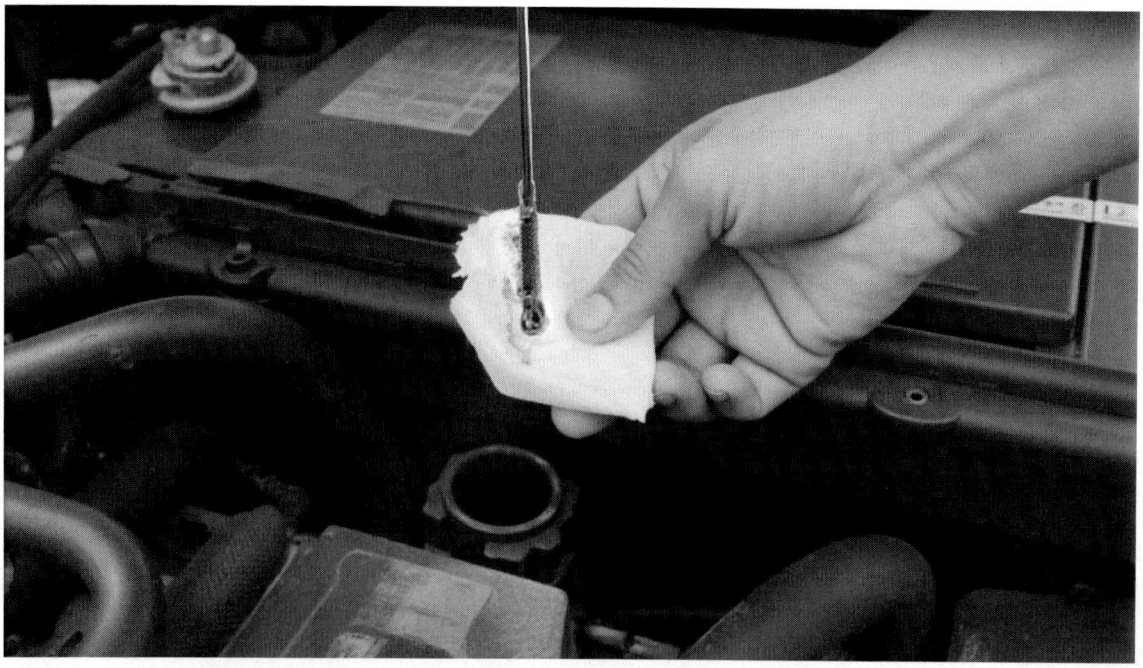

Safety Precautions:

- Always use jack stands when lifting the car—jacks alone are not stable enough.
- Be careful with hot engine components to avoid burns.
- Dispose of old oil responsibly at an auto parts store or recycling center. Never dump it down a drain or in the trash.

Common Pitfalls and Troubleshooting Tips:

- **Over-Tightening the Oil Filter:** It's tempting to crank the oil filter on as tightly as possible, but this can damage the gasket and make future oil changes difficult. Hand-tightening is usually enough.
- **Forgetting to Check for Leaks:** After reattaching the drain plug and filter, always check for leaks by running the engine for a minute. Look underneath the car to ensure no oil is dripping.
- **Using the Wrong Oil:** Not all engine oils are the same. Always use the oil type and viscosity recommended in your car's manual. Using the wrong oil can impact engine performance and fuel efficiency.

Changing your own oil might seem intimidating at first, but once you've done it, you'll realize how straightforward—and satisfying—it is. Not only do you save money, but you also gain the confidence to take better care of your car. Plus, you'll know the job was done right because you did it yourself.

Replacing Engine and Cabin Air Filters

Replacing your engine and cabin air filters is one of the easiest and quickest maintenance tasks you can tackle. Both filters play critical roles in keeping your car running efficiently and you breathing clean air.

The engine air filter prevents dirt, debris, and dust from entering the engine, which could otherwise clog components and reduce fuel efficiency. A clean filter helps your engine breathe better, improving its performance and lifespan.

Meanwhile, the cabin air filter keeps the air inside your car fresh by filtering out dust, pollen, and pollutants. If it's clogged, you'll notice weaker airflow from your vents and might even experience unpleasant odors.

The best part? Both filters are inexpensive, and replacing them yourself can save you the steep markup most repair shops charge.

Tools Needed:

- Screwdriver (if required)
- New engine air filter
- New cabin air filter
- Gloves (optional, but helpful if filters are particularly dirty)

Estimated Time: 10–20 minutes

Instructions:

Replacing the Engine Air Filter

1. **Locate the Air Filter Housing:** Pop the hood and find the air filter housing. It's usually a rectangular or square plastic box near the engine, with clips or screws securing the cover.

2. **Open the Housing:** Unclip or unscrew the cover. Set any screws aside somewhere safe—they're easy to lose.
3. **Remove the Old Filter:** Pull out the old filter. Take a moment to inspect it—if it's clogged with dirt, leaves, or debris, it is definitely time for a replacement.

4. **Install the New Filter:** Place the new filter in the same orientation as the old one. It should fit snugly inside the housing.

5. **Secure the Housing:** Put the cover back on and clip or screw it into place. Double-check that it's secure to avoid any rattling or air leaks.

Replacing the Cabin Air Filter

1. **Locate the Cabin Filter:** Check your manual for its location. Most cabin filters are behind the glove box or under the dashboard.

2. **Access the Filter:** If it's behind the glove box, open it fully. Some glove boxes require you to remove a side clip or two screws for full access—don't worry; this is easy to do.
3. **Remove the Old Filter:** Slide the old filter out carefully. Take note of the airflow direction—there are usually arrows on the filter indicating this.

4. **Install the New Filter:** Insert the new filter, ensuring the airflow arrows point in the correct direction (usually toward the car's cabin).
5. **Reassemble:** If you removed the glove box or other parts, put them back in place. Close everything up and turn on your car to test airflow from the vents.

Safety Precautions:

- Wear gloves if the filters are dirty—dust and debris can be messy.
- Always ensure the engine is off and cool before working under the hood.

Common Pitfalls and Troubleshooting Tips:

- **Forgetting Airflow Direction:** Cabin filters have arrows indicating airflow direction. Installing it backward will reduce efficiency and airflow. Double-check the arrows during installation.
- **Skipping Regular Replacement:** A neglected engine air filter can reduce fuel economy and engine performance, while a dirty cabin filter will leave you breathing stale, dusty air. Replacing these filters every 12,000–15,000 miles (or sooner if you drive in dusty conditions) is key.
- **Misplacing Screws or Clips:** When removing screws or clips from the air filter housing or glove box, keep them in a small container or set them aside where they won't get lost.

Replacing your engine and cabin air filters is one of those tasks that takes barely any time but delivers big benefits. Your engine will run more efficiently, your cabin air will stay cleaner, and you'll save money doing it yourself.

Adjusting Tire Pressure

You might not think much about your tires until there's a problem, but proper tire pressure is crucial for your car's performance, safety, and fuel efficiency. Underinflated tires wear out faster, reduce fuel economy, and can even lead to blowouts. On the flip side, overinflated tires can reduce traction, making it harder to handle the car, especially in wet or slippery conditions.

The good news? Adjusting tire pressure is one of the simplest car maintenance tasks you can do. In just a few minutes, you can ensure your tires last longer, and your car runs smoother.

Tools Needed:

- Tire pressure gauge
- Air compressor (at home, gas station, or portable inflator)

Estimated Time: 10–15 minutes

Instructions:

- **Check the Recommended Pressure:** Start by finding the correct tire pressure for your vehicle. Look for a sticker inside the driver's door frame, or check your owner's manual. The recommended pressure is typically measured in PSI (pounds per square inch) and often differs for front and rear tires. Avoid relying on the pressure listed on the tire itself—this is the maximum pressure, not the optimal one for your car.

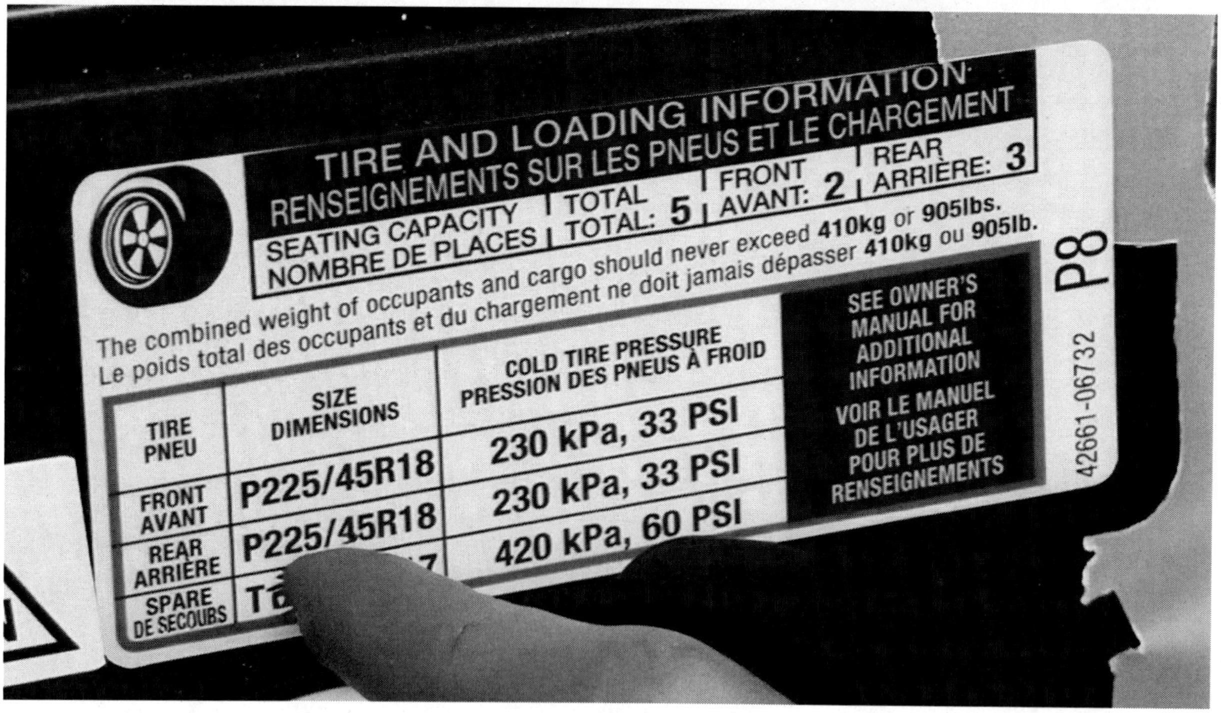

- **Measure Current Pressure:**
  - Remove the valve cap from one tire.
  - Press the tire pressure gauge firmly onto the valve stem until you hear a quick hiss of air and see a reading on the gauge.
  - Note the pressure and compare it to the recommended PSI.
  - Repeat this step for all four tires—and don't forget the spare tire!

- **Add or Release Air:** If the pressure is too low, connect the air compressor to the valve stem and add air in short bursts. Check the pressure frequently to avoid overinflating. If the pressure is too high, press down on the valve pin (you can use the tip of a gauge or your fingernail) to release small amounts of air. Recheck after releasing to ensure it's at the correct PSI.
- **Recheck and Replace Caps:** Once you've adjusted the pressure, double-check each tire to confirm it matches the recommended PSI. Secure the valve caps tightly to prevent dirt and moisture from entering the valve stems.

Safety Precautions:

- Always check tire pressure when the tires are cold—meaning the car hasn't been driven for at least a few hours. Heat causes the air inside the tires to expand, giving a false reading.

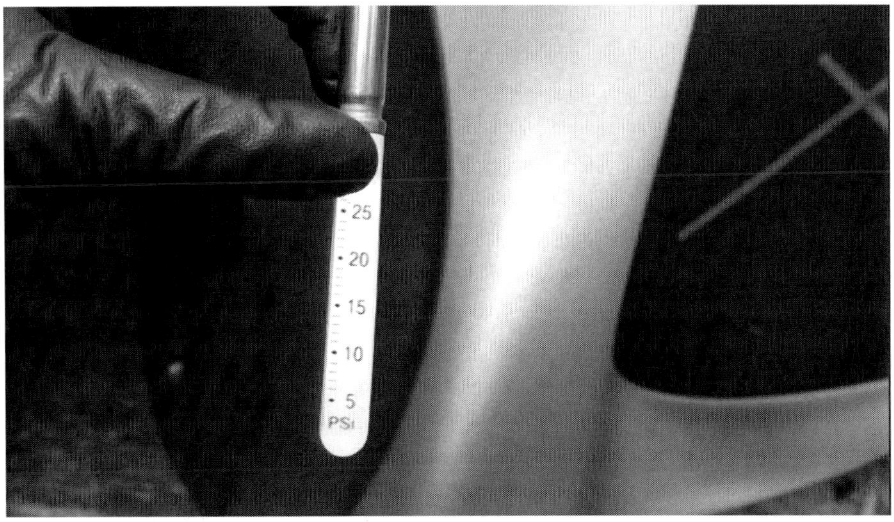

- Avoid overinflating the tires, as this can reduce traction and increase wear on the center of the tread.

Common Pitfalls and Troubleshooting Tips:

- **Forgetting the Spare Tire:** Many drivers overlook the spare tire, but you'll regret it if you need it and it's flat. Make checking the spare part of your routine.
- **Ignoring Seasonal Changes:** Tire pressure drops in cold weather and rises in heat. Be sure to check your tires more frequently during extreme temperature changes.
- **Using Inaccurate Gauges:** Not all tire pressure gauges are equal. Invest in a reliable digital or analog gauge for the most accurate readings. Cheap or worn-out gauges can give misleading results.
- **Assuming Tires Are Fine Without Checking:** Tires can lose air gradually without any visible signs. A quick monthly check is all it takes to avoid surprises on the road.

Proper tire pressure doesn't just save you money on gas and tire replacements—it makes driving safer and more comfortable. Besides, this is a job you can do almost anywhere, whether at home with a portable air compressor or at your local gas station. Take a few minutes to check your tires regularly, and you'll be rewarded with better performance, fewer headaches, and a smoother ride.

Rotating Tires for Even Wear

Tire rotation is one of those simple maintenance tasks that can have a huge impact on your car's performance, safety, and long-term savings. Tires don't wear evenly on all four corners of the car. For example, front tires on front-wheel-drive vehicles wear faster because they handle most of the car's braking, steering, and power delivery.

By regularly rotating your tires, you ensure even wear, extend the life of the tires, and improve handling and traction. What's more, consistent tire rotation can save you money by avoiding the need for premature tire replacement.

Tools Needed:

- Jack and jack stands
- Lug wrench
- Torque wrench (optional but recommended)

Estimated Time: 30–45 minutes

Instructions:

1. Loosen the Lug Nuts

- Before lifting the car, use the lug wrench to slightly loosen the lug nuts on all four tires. Don't fully remove them yet—just break them loose, so they're easier to remove once the car is off the ground.

2. Lift and Secure the Car

- Use the jack to lift one corner of the car at a time and secure it with jack stands. Place the jack stands under the car's designated jack points (check your owner's manual for their location). Never rely on the jack alone to support the car.

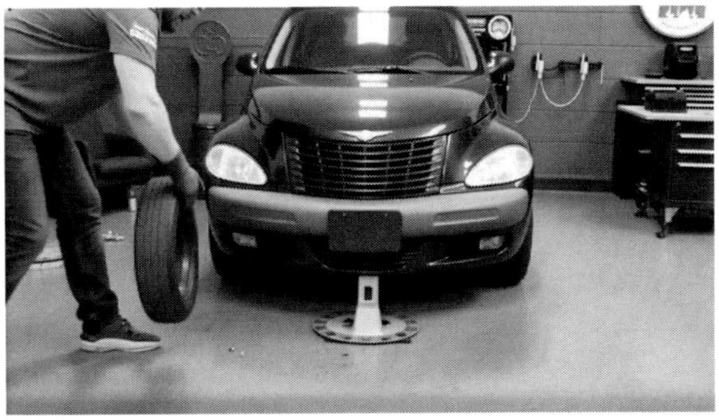

3. Rotate the Tires

- The rotation pattern depends on your tire type:
    - **For non-directional tires (most common):** Move the front tires to the rear and cross the rear tires diagonally to the front. For example, the rear left tire moves to the front right.
    - **For directional tires:** These tires have specific tread patterns and can only roll in one direction. Simply swap the front tires to the rear on the same side.

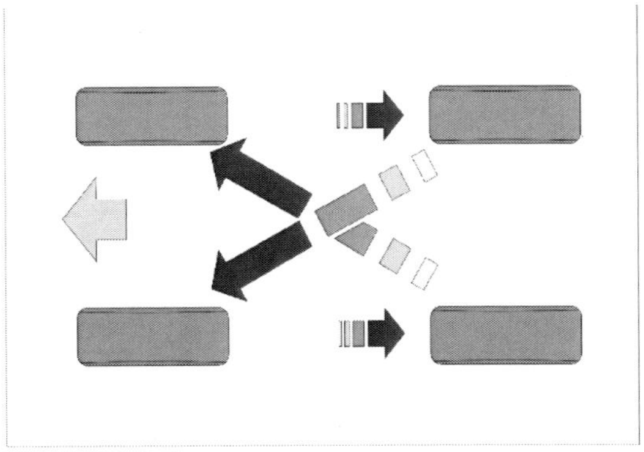

Tip: If you're unsure about your tires, check for arrows or direction markings on the sidewalls.

4. Reattach and Tighten

- Put the tires back onto the wheel hubs and hand-tighten the lug nuts. Once all four tires are in place, lower the car to the ground.
- Use the lug wrench or a torque wrench to tighten the lug nuts fully. Tighten in a star pattern—tightening one nut, then the one opposite it, to evenly secure the wheel. This prevents warping the brake rotors.

Safety Precautions:

- Always work on a flat, stable surface and engage the parking brake.
- Never place any part of your body under the car unless it's securely supported with jack stands.
- Ensure the lug nuts are tightened to the manufacturer's recommended torque specifications. This information can usually be found in your owner's manual.

Common Pitfalls and Troubleshooting Tips:

- **Skipping Regular Rotations:** Many drivers overlook this task, but skipping tire rotations leads to uneven tread wear, reduced handling, and shorter tire life. Aim to rotate your tires every 5,000–7,500 miles or as recommended by your vehicle's manual.
- **Using the Wrong Rotation Pattern:** Using the wrong pattern—especially for directional tires—can result in poor performance or even unsafe driving. Always double-check the tire markings and follow the correct rotation guidelines.
- **Not Checking Tire Pressure:** After rotating your tires, check and adjust the tire pressure according to the manufacturer's specifications. Tire pressure often differs between front and rear tires, so this step ensures optimal performance and safety.
- **Forgetting to Tighten Lug Nuts Properly:** Under-tightened lug nuts can cause the wheel to wobble or come loose, while over-tightened nuts may warp the brake rotors. If possible, use a torque wrench to ensure proper tightening.

Rotating your tires doesn't just save you money by extending their life—it also improves your car's overall performance and safety. You'll enjoy a smoother, more balanced ride and better traction in all conditions. Best of all, it's a quick, simple task you can easily do at home with a few basic tools.

Once you get the hang of it, tire rotations will become a part of your regular maintenance routine, saving you from unexpected tire replacements and costly repairs.

Plugging a Tire Puncture

Discovering a nail or screw in your tire doesn't mean you need to replace it immediately. In most cases, you can repair it yourself and get back on the road quickly. Plugging a tire is a temporary solution for small punctures in the tread, saving you the hassle of a tow or an expensive new tire. However, keep in mind this is only for minor damage—sidewall punctures or large gashes require a full replacement.

Tools Needed:

- Tire repair kit
- Pliers
- Air compressor or tire inflator
- Utility knife
- Soapy water spray bottle

Estimated Time: 15–30 minutes

Instructions:

- **Locate the Puncture:** Inspect the tire to find the source of the leak, like a nail or screw. Listen for a hissing sound if the puncture isn't obvious. Here's a pro tip: spray soapy water on the tire. Bubbles will form at the leak point.
- **Remove the Object:** Use pliers to carefully pull out the embedded object. Take your time—rushing could make the hole larger or cause additional damage.

- **Ream the Hole:** Grab the reaming tool from your tire repair kit. Insert it into the puncture hole and work it in and out a few times. This cleans and roughens the hole to help the plug seal effectively. It may require a bit of elbow grease, but don't be afraid to push firmly.

- **Insert the Plug:** Take one of the sticky rubber plug strips and thread it through the eye of the insertion tool so the ends hang evenly. Push the tool and plug into the hole until only about an inch of the plug remains visible outside the tire. Pull the tool out quickly but carefully, leaving the plug in place. Another pro tip: if it feels tight, that's a good sign. The plug needs to form a solid seal.

- **Trim the Excess:** Use a utility knife to trim off any excess plug sticking out of the tire surface. Leave a small amount flush with the tread to ensure the seal stays secure.

- **Reinflate the Tire and Test:** Use an air compressor or tire inflator to refill the tire to its recommended pressure (check the driver's side door sticker for the correct PSI). To confirm the repair worked, spray soapy water over the plugged area. If no bubbles form, the leak is sealed.

Safety Precautions:

- Never repair a sidewall puncture. The sidewall of the tire is under constant stress as you drive, and a plug won't provide sufficient strength to hold up under these conditions. Driving on a damaged sidewall can result in a blowout, so always replace the tire instead.
- Avoid driving on a flat tire. This can damage the tire's structure, making it unsafe to repair.
- Check your tire pressure regularly. A repaired tire can slowly lose pressure over time, so keep an eye on it.

Common Pitfalls and Troubleshooting Tips:

- **Failing to Properly Ream the Hole:** If you skip reaming or don't do it thoroughly, the plug may not seal properly, and the tire will continue to leak. Take a few extra moments to prepare the hole.
- **Not Checking for Leaks:** After the repair, always test with soapy water to confirm the plug is airtight. Missing this step can leave you with a slow leak that becomes problematic later.
- **Plugging Unrepairable Damage:** If the hole is larger than ¼ inch or in the tire's sidewall, don't attempt a repair. It's safer to replace the tire.

Plugging a tire might seem intimidating at first, but it's a straightforward process that can get you back on the road quickly. It's also an invaluable skill to have, especially during emergencies or road trips. Once you've done it a couple of times, you'll wonder why you ever considered calling for help. Keep a tire repair kit and a portable air compressor in your trunk—they'll come in handy when you least expect it. And remember, while a plug is a great temporary fix, it's still a good idea to have the tire inspected or patched professionally for long-term safety.

Replacing Windshield Wiper Blades

Windshield wipers are easy to overlook—until you're caught in a downpour with streaky, ineffective blades. Clear visibility is critical for safe driving, and replacing worn-out wiper blades is one of the easiest and most inexpensive maintenance tasks you can do. Best of all, you don't need any tools or mechanical experience to tackle this quick fix.

Tools Needed:

- New windshield wiper blades (check your manual or measure the existing blades for size)

Estimated Time: 5–10 minutes

Instructions:

- **Lift the Wiper Arms:** Start by gently lifting the wiper arms away from the windshield until they're upright. Be careful not to let the arms snap back, as they can crack or damage the glass.
- **Remove the Old Wiper Blades:** Locate the release tab on the wiper blade (usually where the blade connects to the arm). Press or squeeze the tab, then slide the old blade downward to detach it from the hook or clip. Some blades require you to rotate them slightly to slide them off. If you're struggling, refer to the new blades' instructions for guidance—many manufacturers include a diagram. If you're working in cold weather, the rubber on old blades can become stiff, making removal trickier. A little patience goes a long way!

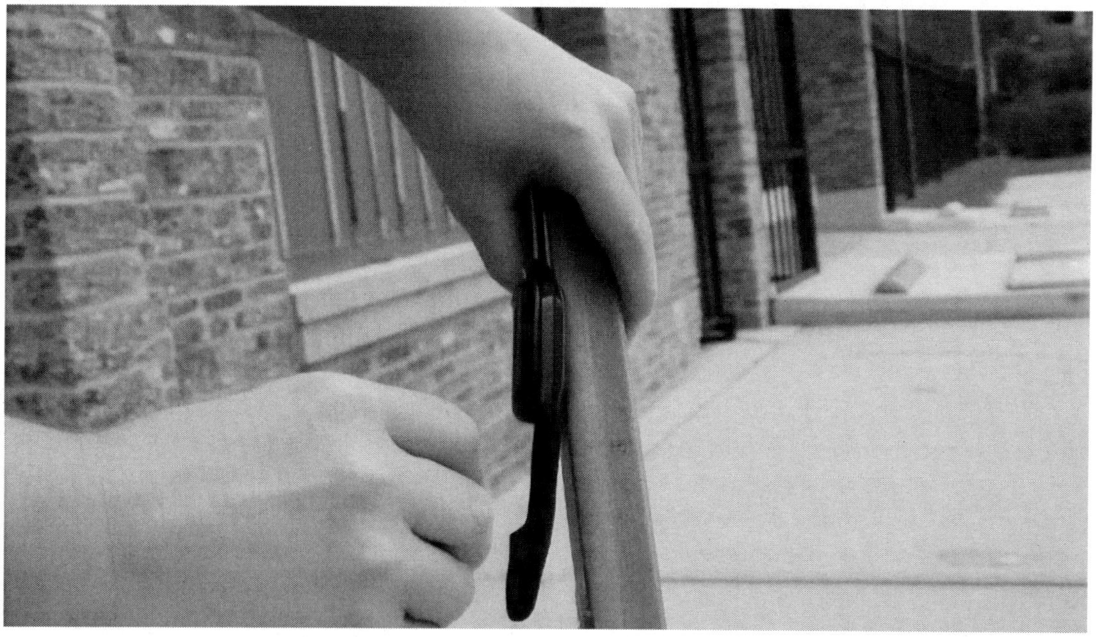

- **Attach the New Wiper Blades:** Align the new wiper blade with the wiper arm's hook or clip. Slide it into place until you hear a firm "click," which indicates it's locked in securely. Give it a gentle tug to ensure it won't come loose. Repeat the process for the other blade.
- **Test the Wipers:** Carefully lower the wiper arms back onto the windshield. Start the car and test the wipers to ensure they move smoothly without skipping or making noise. Check for any streaks on the windshield; if you see any, confirm the blades are installed correctly.

If you live in a cold climate, check your wipers regularly during winter months to ensure they haven't frozen to the windshield. Frozen wipers can tear or crack when you turn them on, so it's essential to keep them free of ice and frost.

Safety Precautions:

- Be gentle when lifting the wiper arms. They're spring-loaded, and accidentally letting go can cause them to snap back, potentially damaging the windshield.
- If you're working with a tall or large vehicle, use a step stool to avoid straining yourself.
- Always handle the blades by the frame, not the rubber, to avoid bending or damaging them.

Common Pitfalls and Troubleshooting Tips:

- **Not Checking the Blade Size:** Wiper blades come in various lengths, and it's not always one-size-fits-all. Double-check your owner's manual or measure your current blades to ensure you buy the correct size. Some cars use different lengths for the driver and passenger sides, so don't assume they're identical.
- **Installing the Blade Incorrectly:** If the blade isn't fully locked into place, it can come loose while driving—an issue you don't want to discover during heavy rain. Always listen for the "click" when attaching the new blade and test it afterward.
- **Forgetting the Rear Wiper:** If your car has a rear windshield wiper, check it, too! Rear wiper blades often get neglected but are just as important for visibility in bad weather.

Replacing windshield wiper blades is one of those quick wins in car maintenance. It's easy, requires no tools, and takes only a few minutes. Plus, the payoff is huge: better visibility, safer driving, and peace of mind during rain or snow.

Keep an eye on your blades, especially if they start streaking, squeaking, or skipping across the glass. Most wiper blades last 6–12 months, so it's a good habit to replace them at least once a year or whenever you notice signs of wear.

Cleaning Battery Terminals

Corroded battery terminals might seem like a small issue, but they can lead to big problems, like starting difficulties or power loss. Over time, the buildup of whitish or greenish residue on the terminals interferes with the battery's ability to deliver electricity efficiently. Cleaning them is a quick and straightforward way to keep your car running smoothly.

Tools Needed:

- Adjustable wrench or socket wrench
- Baking soda
- Water
- Old toothbrush or wire brush

- Clean rag or paper towels
- Petroleum jelly (optional)
- Rubber gloves and safety goggles

Estimated Time: 15–20 minutes

Instructions:

- **Turn Off the Car:** Start by ensuring the engine is off, the car is in park, and the keys are out of the ignition. This step is essential for your safety.
- **Disconnect the Battery:** Always begin by disconnecting the negative terminal (marked with a minus "–"). This reduces the risk of sparks or short circuits. Loosen the nut using a wrench and lift the terminal off. Next, disconnect the positive terminal (marked with a plus "+") in the same manner.

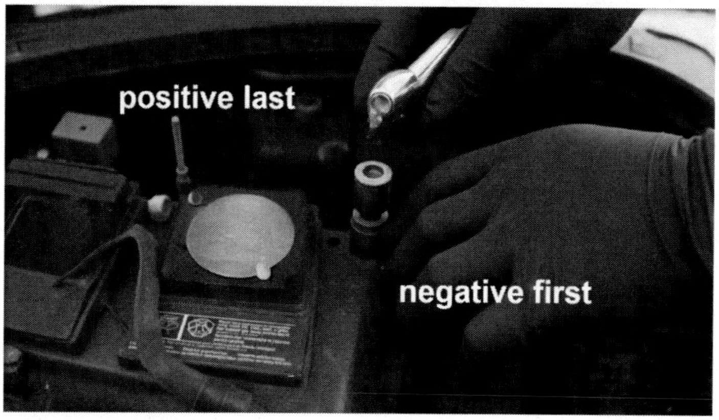

- **Prepare a Cleaning Solution:** Mix one tablespoon of baking soda with a cup of water. This mild solution neutralizes corrosion without damaging the terminals.
- **Scrub the Terminals:** Put on your gloves and goggles. Use an old toothbrush or wire brush dipped in the baking soda solution to scrub the terminals and clamps. Focus on any areas with visible corrosion. Don't forget to clean the terminal posts on the battery and the inside of the clamps.

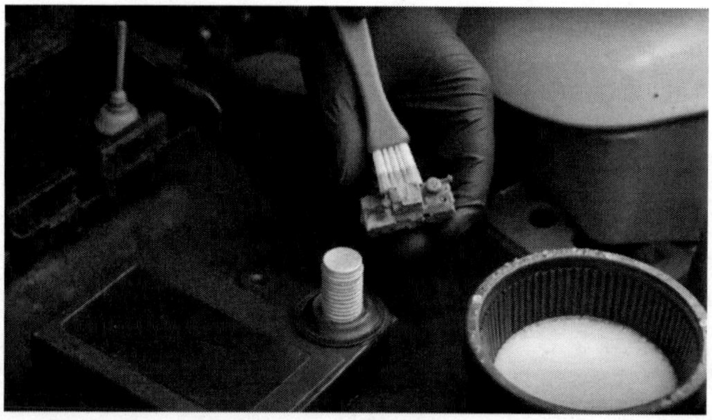

- **Rinse and Dry:** Wipe away the baking soda residue with a rag dipped in clean water. Ensure no residue is left, as it can cause future buildup. Dry everything thoroughly with a clean rag or paper towel. Moisture can lead to corrosion or electrical issues.
- **Reconnect the Battery:** Reattach the positive terminal first, tightening it snugly with a wrench. Then, reconnect the negative terminal and tighten it securely. Avoid overtightening, as it can damage the clamps.

- **Protect the Terminals:** Apply a thin layer of petroleum jelly or a battery terminal protectant spray to help prevent future corrosion. This step adds a layer of protection against moisture and air exposure.

Safety Precautions:

- Wear gloves and goggles. Battery acid and corrosion can irritate your skin and eyes. Always protect yourself.
- Always disconnect the negative terminal first and reconnect it last to minimize the risk of short circuits.
- Work in a ventilated area. Batteries can emit gases, so choose an open space to avoid inhaling fumes.

Common Pitfalls and Troubleshooting Tips:

- **Not drying thoroughly:** Moisture left behind can accelerate corrosion. Double-check that the terminals and clamps are completely dry before reconnecting.
- **Loose connections:** If the clamps aren't tightened securely, your car may struggle to start or experience electrical issues. Make sure they're snug, but avoid overtightening.
- **Recurring corrosion:** If corrosion keeps returning despite cleaning, it may indicate a deeper issue, like a failing battery or overcharging alternator. Consider having your battery tested to rule out underlying problems.

Cleaning your battery terminals is a small but impactful task that improves your car's electrical performance and extends the life of your battery. It's quick, easy, and requires just a few household items. By making this a regular part of your car maintenance routine, you'll save yourself from unnecessary starting issues and costly repairs down the line.

Fixing a Faulty Headlight or Taillight Bulb

Your headlights and taillights are essential for safety—not just yours but everyone else's on the road. A faulty light reduces your visibility and makes it harder for others to see your car, especially at night or in bad weather. Thankfully, replacing a bulb is one of the simplest DIY car repairs, and it saves you the cost of labor at a repair shop.

Tools Needed:

- Replacement bulb (match the part number in your manual or on the old bulb)
- Screwdriver (if required for your vehicle)
- Clean gloves or a lint-free rag

Estimated Time: 10–20 minutes

Instructions:

- **Identify the Correct Bulb:** Check your owner's manual or inspect the faulty bulb to find the correct replacement. Make sure the wattage and part number match, as using the wrong bulb can damage your car's electrical system.
- **Access the Bulb Housing:**
    - **Headlights:** Pop the hood and locate the back of the headlight assembly. Depending on your car, you may need to remove a panel or unclip parts to access the bulb.

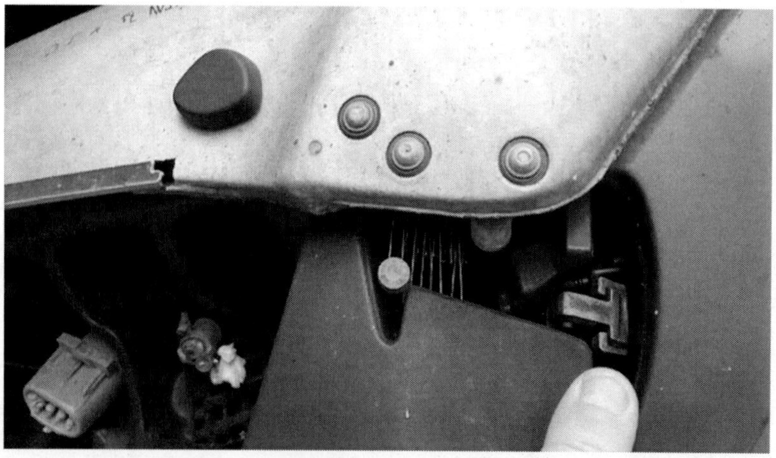

- **Taillights:** Open the trunk and find the bulb housing, which may be behind a panel inside the car or directly behind the taillight lens.

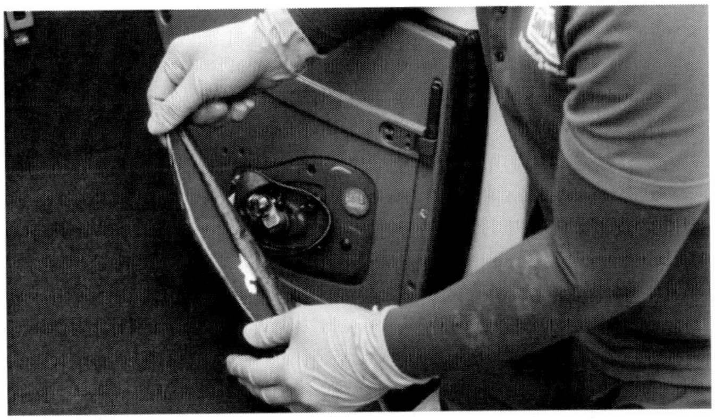

- **Remove the Old Bulb:** Turn the bulb socket counterclockwise to unlock it, then carefully pull the bulb out. If you're replacing a halogen bulb, avoid touching the glass directly. Oil from your skin can create hot spots on the bulb, shortening its life. Use gloves or a clean rag for handling.

- **Install the New Bulb:** Insert the new bulb into the socket, ensuring it's aligned correctly, and twist it clockwise to lock it securely in place. Double-check that it's seated properly—loose bulbs may not function or could flicker.

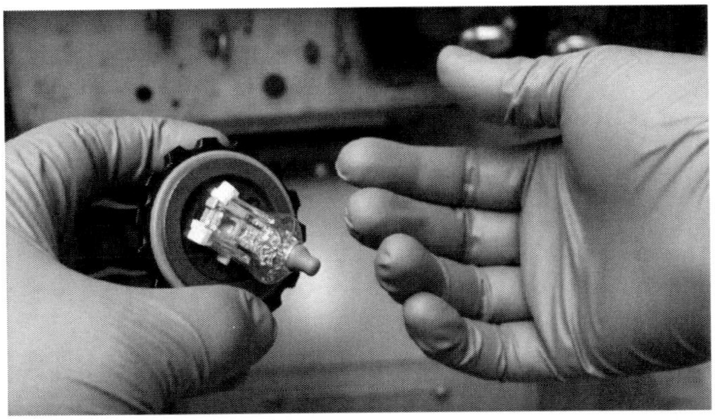

- **Test the Light:** Before reassembling any panels, turn on the headlights or taillights to ensure the new bulb works. This step saves you the hassle of redoing the process if something isn't right.
- **Reassemble the Housing:** Replace any panels, clips, or screws you removed earlier. Close the hood or trunk, and you're done!

Safety Precautions:

- Always turn off the car and lights before starting. Working on live bulbs can cause burns or electrical shorts.
- Use gloves or a clean rag to handle bulbs, especially halogen ones, to avoid leaving oil residue.

Common Pitfalls and Troubleshooting Tips:

- **Using the wrong bulb:** Double-check the part number. If you're unsure, bring the old bulb to the store for comparison.
- **Loose connections:** If the bulb doesn't work after replacement, make sure it's seated correctly and that the socket isn't damaged.
- **Still not working?** The problem might be a blown fuse or a wiring issue. Use a multimeter to test for power at the socket, or consult your manual to check the fuse associated with the faulty light.

INTERMEDIATE PROJECTS: STEP UP YOUR DIY GAME

Once you're comfortable with the basics, these tasks will help you develop new skills and save even more money.

Replacing Brake Pads

Your brakes are one of the most critical safety systems in your car. Over time, brake pads wear down, reducing their ability to stop the vehicle effectively. Replacing them before they're completely worn out ensures your car stops safely and prevents costly damage to the brake rotors. The process might seem intimidating, but with the right tools and guidance, you can do it yourself and save a significant amount of money.

Tools Needed:

- Lug wrench
- Jack and jack stands
- Socket wrench
- C-clamp or brake caliper tool
- New brake pads

- Brake grease
- Gloves and safety goggles

Estimated Time: 45–60 minutes

Instructions:

- **Park Safely and Lift the Car:** Find a flat, stable surface to park your car. Engage the parking brake and place wheel chocks behind the rear tires. Loosen the lug nuts on the wheel you'll be working on, but don't remove them yet. Use the jack to lift the car and secure it with jack stands. Then, remove the wheel entirely to access the brake assembly.
- **Access the Brake Pads:** Locate the brake caliper, which is the part that houses the brake pads. Use a socket wrench to remove the bolts holding the caliper in place. Carefully slide the caliper off, ensuring you don't let it hang by the brake line. Rest it on a box or secure it with a bungee cord.

- **Remove the Old Pads:** Slide out the old brake pads, noting their position. This will help you install the new ones correctly later. Check the rotor for any deep grooves or damage that may indicate it needs resurfacing or replacement.

- **Compress the Caliper Piston:** The caliper piston needs to be pushed back to accommodate the thickness of the new brake pads. Use a C-clamp or brake caliper tool to slowly and evenly compress the piston back into the caliper housing. Be patient to avoid damaging the piston.

- **Install the New Pads:** Apply a thin layer of brake grease to the back of the new pads—avoid getting any on the friction surface. Slide the new pads into place, ensuring they fit snugly.

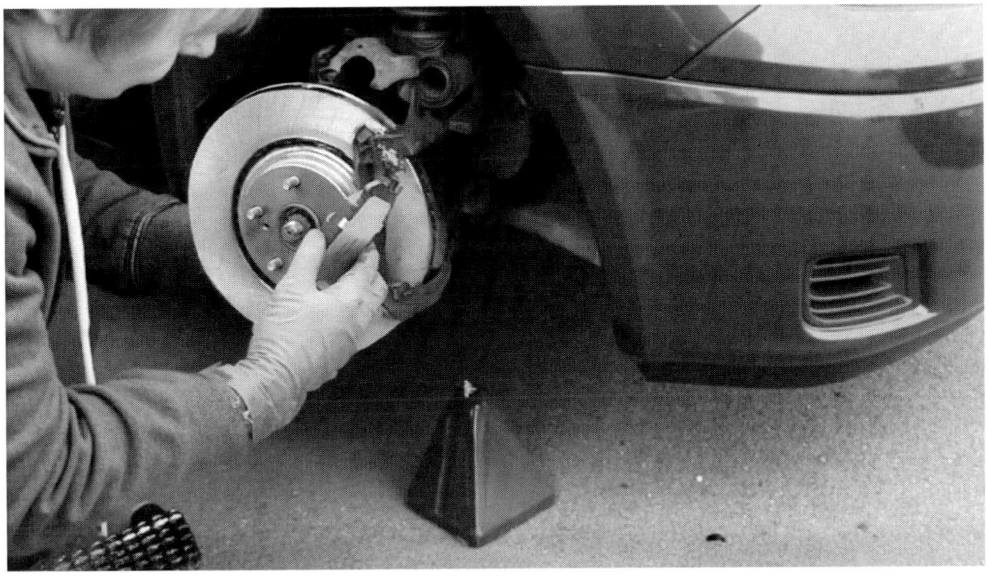

- **Reassemble the Caliper:** Place the caliper back over the new pads and reattach the bolts. Tighten them securely with your socket wrench.

- **Repeat for the Other Side:** Always replace brake pads in pairs to maintain balanced braking. Repeat the process for the other wheel on the same axle.
- **Test the Brakes:** Before driving, pump the brake pedal several times until it feels firm. This step restores the hydraulic pressure and ensures your brakes are ready to use.

Safety Precautions:

- **Always use jack stands:** Never trust the jack alone to hold the car.
- **Avoid touching critical surfaces:** Keep the friction surface of the pads and the rotor clean to maintain braking efficiency.
- **Wear protective gear:** Gloves and goggles protect you from brake dust and grease.

Common Pitfalls and Troubleshooting Tips:

- **Forgetting to pump the brake pedal:** If you skip this step, your brakes may not engage when you first drive. Always test them before hitting the road.
- **Improper lubrication:** If the brakes squeal after replacement, double-check that you applied brake grease to the pad backs—not the friction surface.
- **Caliper alignment:** Ensure the caliper is seated properly before tightening the bolts. Misalignment can cause uneven pad wear or reduced braking efficiency.

Replacing brake pads might sound intimidating, but with a little preparation and patience, it's a manageable DIY project. Not only will you save on labor costs, but you'll also gain the satisfaction of knowing your car is safer because of your work.

Flushing Brake Fluid

Brake fluid plays a critical role in your car's braking system, transferring force from your foot on the pedal to the brake calipers and pads. Over time, brake fluid absorbs moisture, which can reduce braking efficiency and cause corrosion in the system. Flushing and replacing the brake fluid is a straightforward DIY task that improves performance and ensures safety. Here's how to do it:

Tools Needed:

- Wrench set
- Brake fluid (consult your manual for the correct type)
- Brake bleeder kit or clear tubing and a container
- Jack and jack stands
- Gloves and safety goggles

Estimated Time: 45 minutes

Instructions:

- **Prepare the Car:** Park your car on a flat, level surface and engage the parking brake. Use a jack to lift the car and secure it with jack stands for stability. This will give you easy access to all four wheels.
- **Locate the Brake Fluid Reservoir:** Pop the hood and locate the brake fluid reservoir, typically near the back of the engine bay on the driver's side. Wipe the cap and surrounding area clean to prevent dirt from contaminating the system when you open it.

- **Remove Old Fluid:** Using a turkey baster or hand pump, siphon out as much old brake fluid from the reservoir as you can. Avoid touching the fluid directly, as it's corrosive. Once the reservoir is mostly empty, refill it with fresh brake fluid to the recommended level.

- **Bleed the Brakes:** Bleeding the brakes ensures that old fluid and any trapped air are removed.
    - **Start at the Farthest Wheel:** Begin with the wheel farthest from the reservoir (usually the rear passenger side). Attach clear tubing to the brake bleeder valve, with the other end placed into a container to catch the old fluid.

    - **Have a Helper Pump the Brakes:** Ask a helper to press and hold the brake pedal. While they do this, loosen the bleeder valve with a wrench to allow old fluid to flow out.
    - **Tighten the Valve:** Once the fluid flow stops, tighten the valve and have your helper release the pedal.
    - **Repeat Until Clear Fluid Flows:** Continue the process until fresh, clean fluid comes through the tubing.
- **Move to the Next Wheel:** Repeat the bleeding process, moving to the next farthest wheel and working your way closer to the reservoir. Check the fluid level in the reservoir frequently and top it off as needed to prevent air from entering the system.
- **Test the Brakes:** Once all four wheels are done, pump the brake pedal a few times to restore pressure. Take the car for a slow, cautious test drive to ensure the brakes feel firm and responsive.

Safety Precautions:

- **Wear gloves and goggles:** Brake fluid is corrosive and can irritate skin and eyes.
- **Avoid spills:** Wipe up any spilled brake fluid immediately to prevent damage to painted surfaces.
- **Keep the reservoir topped off:** Allowing it to run dry can introduce air into the system, leading to spongy brakes.

Common Pitfalls and Troubleshooting Tips:

- **Air in the system:** If the brakes feel spongy after flushing, it means air is trapped in the lines. Re-bleed the brakes to remove all air bubbles. Sometimes, it may take more than one round of bleeding to fully eliminate stubborn air pockets.
- **Using the wrong fluid:** Always use the type of brake fluid specified in your owner's manual. Never mix different types of brake fluid, such as DOT 3 and DOT 4, as they have different chemical compositions and boiling points. Mixing fluids can lead to reduced braking efficiency, compromised system performance, and even potential damage to the braking system.
- **Skipping regular flushes:** Brake fluid should be replaced every 2–3 years to maintain performance. Neglecting this can lead to reduced braking efficiency or system damage.
- **Improper disposal of old fluid:** Brake fluid is classified as hazardous waste and should be disposed of properly. Check with your local recycling center or hazardous waste facility for safe disposal options.

Flushing your brake fluid is a rewarding task that enhances your car's safety and braking power. It might sound intimidating at first, but with the right tools, a little patience, and this guide, you can tackle it with confidence. Keeping your brake system in top shape isn't just about performance—it's about peace of mind every time you're behind the wheel.

Fixing a Flat Tire and Installing a Spare

A flat tire can be an unwelcome surprise, but with a little preparation and know-how, you can get back on the road in no time. Changing a flat tire and installing a spare is a straightforward process that every car owner should know. Here's your step-by-step guide:

Tools Needed:

- Spare tire
- Lug wrench
- Jack and jack stands
- Wheel chocks

Estimated Time: 15–30 minutes

Instructions:

- **Park Safely:** Your safety comes first. Pull over to a flat, stable location away from traffic. Turn on your hazard lights, engage the parking brake, and place wheel chocks on the wheels opposite the flat tire to prevent movement.

- **Loosen the Lug Nuts:** Use your lug wrench to loosen the lug nuts on the flat tire. Turn them counterclockwise, but don't remove them completely yet. Breaking them loose while the car is still on the ground prevents the tire from spinning.

- **Lift the Car:** Position your jack under the designated jacking point nearest the flat tire. Most vehicles have specific jacking points to ensure stability and prevent damage—check your owner's manual to locate them. Use the jack to lift the car until the flat tire is fully off the ground, and then secure the car with jack stands for added safety. It's critical to ensure the car is securely elevated before proceeding.

- **Remove the Flat Tire:** Fully remove the loosened lug nuts and store them in a safe spot. Slide the flat tire off the wheel hub and place it flat on the ground to avoid it rolling away.

- **Install the Spare:** Align the spare tire's holes with the wheel bolts, then slide it onto the hub. Hand-tighten the lug nuts onto the bolts to hold the tire in place.
- **Lower the Car:** Use the jack to slowly lower the car until the spare tire touches the ground and the car's weight is resting on it. Remove the jack and tighten the lug nuts fully in a star pattern (tighten one nut, then the one opposite, and so on). This ensures even pressure on the wheel.
- **Double-Check and Test:** Ensure all lug nuts are secure. Drive at low speeds for a short distance to confirm the spare tire is stable.

Safety Precautions:

- **Use Wheel Chocks:** Always secure the opposite wheels to prevent the car from rolling.
- **Use the Correct Jacking Point:** Always consult your owner's manual to locate the proper jacking points for your vehicle. Improper placement can damage the car or cause instability.
- **Ensure Secure Elevation:** Never work on a car supported solely by a jack. Always use jack stands for added safety.
- **Check the Spare's Condition:** Make sure your spare tire is properly inflated and in good condition before installation.

Common Pitfalls and Troubleshooting Tips:

- **Stuck Lug Nuts:** Lug nuts can be stubborn. If they won't budge, use your body weight or a breaker bar for extra leverage. A quick stomp on the lug wrench (with caution) can also help loosen them.
- **Incorrect Spare Tire Usage:** Remember, spare tires—especially smaller "donut" spares—are temporary solutions. They're not designed for extended use or high speeds. Repair or replace your flat tire as soon as possible.
- **Forgetting the Star Pattern:** Tightening lug nuts randomly can lead to uneven pressure and wobbling. Always tighten them in a star pattern to distribute the force evenly.

Fixing a flat tire might feel daunting at first, but with these clear steps and a little practice, you'll handle it like a pro. Having the tools on hand and knowing how to use them can save you time, money, and frustration. Most importantly, it empowers you to stay safe and get back on the road quickly.

Cleaning and Restoring Cloudy Headlights

Foggy headlights aren't just an eyesore—they're also a safety concern. Cloudy lenses can significantly reduce your nighttime visibility, making it harder to see the road and for other drivers to see you. Thankfully, restoring your headlights is a straightforward project that doesn't require fancy tools or much time. Here's how you can make your headlights look brand new:

Tools Needed:

- Headlight restoration kit (or baking soda and water)
- Microfiber cloths
- Water
- Masking tape (optional but recommended)

Estimated Time: 30–45 minutes

Instructions:

- **Clean the Surface:** Start by washing the headlights with soapy water to remove dirt, grime, and debris. This step ensures you're working with a clean surface. Once clean, dry the headlights thoroughly. Use a microfiber cloth to ensure they are completely free of moisture before moving to the next step. This ensures no residual moisture affects the polishing process.
- **Mask the Surrounding Area:** Use masking tape to cover the edges of your headlights where they meet the paint. This step is especially important if you're sanding or using a polishing compound, as it prevents accidental scratches on your car's finish.

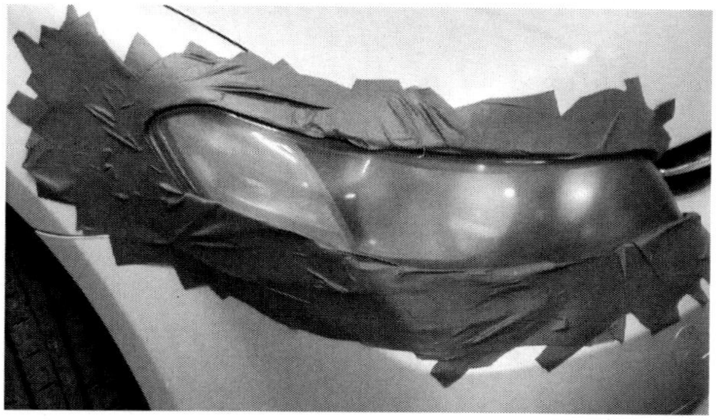

- **Polish the Headlights:** Now it's time to tackle the cloudiness. Choose one of the following methods based on what you have on hand:
  - **Using a Headlight Restoration Kit:** Most kits come with sanding pads and a polishing compound. Follow the kit's instructions, sanding in a consistent motion and applying polish as directed.

  - **DIY Method with Baking Soda and Water:** Mix baking soda with water to form a thick paste. Apply the paste to the lens using a microfiber cloth, rubbing in circular motions. Take your time, as gentle pressure and consistent motion are key to removing the cloudiness.
- **Rinse and Dry:** Once you're satisfied with the polishing, wipe away any residue using a clean, damp microfiber cloth. Rinse the headlights with water and dry them thoroughly.
- **Apply a Sealant:** To protect your hard work, apply a headlight sealant or a layer of car wax to the lens. This step creates a protective barrier that prevents future cloudiness caused by UV exposure and environmental factors.

Safety Precautions:

- **Protect Your Paint:** Always use masking tape to shield the edges of your headlights. This simple step avoids accidental damage to your car's paint job.
- **Apply Even Pressure:** Whether you're sanding or polishing, use consistent, gentle pressure to avoid scratching or uneven results.

Common Pitfalls and Troubleshooting Tips:

- **Skipping the Sealant:** One of the most common mistakes is skipping the sealant step after polishing. Without protection, your headlights will cloud up again quickly, undoing all your hard work.
- **Rushing the Process:** Polishing takes time. Don't rush! Spend a few extra minutes to ensure you've removed as much cloudiness as possible. Hurrying can leave you with uneven results.
- **Using the Wrong Materials:** Stick to the recommended tools and solutions. Abrasive materials like kitchen sponges or harsh chemicals can cause more damage than good.

Restoring cloudy headlights is one of the easiest and most satisfying car maintenance tasks you can tackle. Not only does it improve your car's appearance, but it also enhances your safety on the road. With just a little effort and the right tools, you'll be amazed at the difference clear headlights can make.

Replacing a Serpentine Belt

When it comes to keeping your engine accessories running smoothly—like the alternator, power steering pump, and air conditioning—the serpentine belt does the heavy lifting. Replacing it might seem intimidating, but with the right tools and some patience, it's a task you can tackle at home.

Tools Needed

- Wrench set or serpentine belt tool
- New serpentine belt (double-check your car's manual to confirm the correct part for your car)
- Socket wrench (optional)

Estimated Time: 30–45 minutes

Step-by-Step Instructions

- **Locate the Belt Diagram:** Before you start, locate the routing diagram for your serpentine belt. It's often on a sticker under the hood or in your owner's manual. If it's missing, snap a picture or draw a quick sketch to guide you during installation.

- **Release Belt Tension:** Find the belt tensioner pulley, which keeps the serpentine belt tight. Use a wrench or serpentine belt tool to rotate the tensioner, relieving the tension on the belt. Keep steady pressure while sliding the belt off one pulley—this makes removal easier.

- **Remove the Old Belt:** With the tension released, slide the belt off the remaining pulleys and lift it out of the engine bay. Take a moment to inspect the pulleys for wear or damage.

- **Install the New Belt:** Position the new serpentine belt, threading it over the pulleys in the exact order shown on the diagram. Leave the tensioner pulley for last, as it simplifies the process.

- **Reapply Tension:** Rotate the tensioner again and slip the belt over its pulley. Slowly release the tensioner, allowing it to tighten the belt into place.

- **Inspect and Test:** Double-check that the belt is seated properly on all pulleys and follows the correct routing. Start the engine briefly to ensure the belt operates smoothly without slipping or making noise.

Safety Precautions

- Always work on a cool engine to avoid burns.
- Use caution when releasing or reapplying tension—don't let the wrench snap back suddenly.

Common Pitfalls and Troubleshooting Tips

- **Belt Slipping Off During Installation**: If the belt keeps slipping, check your routing diagram for accuracy. Misalignment on even one pulley can cause issues.
- **Squealing or Chirping Noises**: This often means the belt isn't aligned properly or the tensioner is loose. Double-check the installation and ensure the tensioner is functioning correctly.
- **Old Tensioner Issues**: A worn-out tensioner may not hold proper tension. If you notice slack in the belt, consider replacing the tensioner.

Replacing a serpentine belt is one of those repairs that's surprisingly straightforward once you've done it. Not only will you save money by avoiding the shop, but you'll also gain confidence in tackling more advanced DIY projects. Plus, a properly installed belt keeps your car running reliably, so it's worth every minute you spend on this task!

Cleaning the Mass Airflow Sensor

Keeping your mass airflow (MAF) sensor clean is one of the easiest ways to improve your car's performance and fuel efficiency. The MAF sensor measures the amount of air entering the engine, helping your vehicle adjust the fuel-to-air ratio for optimal combustion. Over time, this sensor can get clogged with dirt and debris, leading to poor acceleration, reduced gas mileage, or even a check engine light. Cleaning it is a quick and cost-effective fix that you can handle at home.

Tools Needed:

- Mass Airflow Sensor Cleaner (do NOT use carburetor or brake cleaner)
- Screwdriver (Phillips or flathead, depending on your car)
- Safety gloves and goggles

Estimated Time: 15–20 minutes

Instructions:

- **Locate the MAF Sensor:** Pop the hood and locate the MAF sensor. It's usually found between the air filter housing and the engine's intake manifold. If you're unsure, consult your owner's manual.

- **Remove the Sensor:** Carefully disconnect the electrical connector connected to the sensor. Use a screwdriver to loosen any screws or clamps holding the sensor in place. Gently pull the sensor out, ensuring you don't damage its delicate wires.

- **Spray the Cleaner:** Hold the MAF sensor steady and spray the MAF cleaner directly onto its wires or plate. Follow the cleaner's instructions for proper use—typically, a few short bursts will suffice. Avoid touching the wires, as they're fragile and can break easily. Allow the cleaner to evaporate completely before reinstalling the sensor.

- **Reinstall the Sensor:** Once the sensor is dry, slide it back into place and secure it with screws or clamps. Reconnect the electrical plug, ensuring it clicks into place.

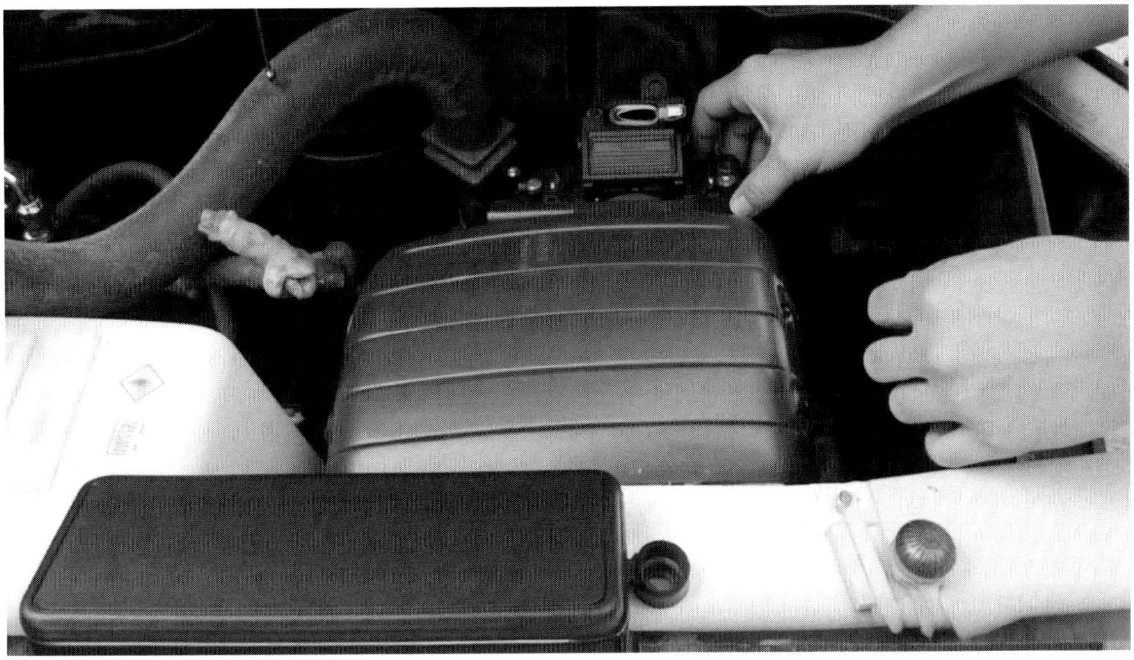

Safety Precautions:

- Always ensure the engine is off, and disconnect the battery to avoid electrical damage.
- Wear gloves and goggles to protect against spray-back or accidental contact with the cleaner.
- Never touch the sensor's internal wires—they are extremely delicate and expensive to replace.

Common Pitfalls and Troubleshooting Tips:

- **Over-Spraying**: Too much cleaner can pool inside the sensor, causing issues. A few short bursts are all you need.
- **Loose Connections**: Double-check that the sensor and its electrical connector are secured properly after cleaning. Loose connections can cause error codes or performance issues.
- **Persistent Check Engine Light**: If the light stays on after cleaning, use an OBD-II scanner to check for specific error codes. This could indicate a deeper issue, such as a failing sensor or other engine-related problems.

A dirty MAF sensor can throw off your car's air-to-fuel ratio, leading to reduced power and fuel inefficiency. Regular cleaning (every 15,000–20,000 miles or as part of your routine maintenance) can extend the sensor's life and keep your car running smoothly. Plus, the process is straightforward and inexpensive compared to the cost of replacing the sensor.

Replacing Spark Plugs

Spark plugs are small but mighty components that play a critical role in your car's performance. They create the spark that ignites the air-fuel mixture in your engine, powering your vehicle. Over time, spark plugs wear out, causing rough idling, poor fuel efficiency, and even engine misfires. Fortunately, replacing them is a straightforward task that can save you a trip to the mechanic.

Tools Needed:

- Socket wrench with a spark plug socket
- Torque wrench
- New spark plugs (check your manual for the correct type)
- Anti-seize lubricant (optional)
- Gap gauge

Estimated Time: 45–60 minutes

Instructions:

- **Locate the Spark Plugs:** Open the hood and identify the spark plugs. They're typically connected to thick wires or covered by ignition coils, depending on your car's design. Your owner's manual will help pinpoint their location.

- **Remove the Old Spark Plugs:** Start with one spark plug at a time to avoid mixing up wires or coils. Carefully disconnect the wire or remove the ignition coil. Use your socket wrench with a spark plug socket to unscrew and gently remove the old spark plug.

- **Check the Gap:** Before installing the new spark plugs, use a gap gauge to verify the gap matches the manufacturer's specifications (found in your manual). If the gap is incorrect, gently adjust it using the gauge.
- **Install the New Spark Plugs:** Hand-tighten the spark plug into place, then use your wrench to secure it snugly. Be careful not to overtighten—it can strip the threads or damage the engine.
- **Reconnect the Wire or Coil:** Reattach the spark plug wire or ignition coil securely. Repeat the process for each remaining spark plug.
- **Test Your Work:** Once all plugs are replaced, start the engine to ensure it runs smoothly. Listen for any misfires or unusual noises.

Safety Precautions:

- Always work on a cold engine to avoid burns.
- Replace spark plugs one at a time to keep the wires and coils organized.
- Wear gloves to protect your hands from dirt and grime.

Common Pitfalls and Troubleshooting Tips:

- **Overtightening Spark Plugs:** Tightening the plugs too much can strip the threads or crack the plug. Use a torque wrench to apply the recommended amount of force.
- **Misfiring Engine:** If the engine feels rough or misfires after replacement, double-check that each spark plug is properly connected and securely tightened. A loose connection or incorrect installation can cause misfires.
- **Wrong Spark Plug Type:** Using the wrong type of spark plug for your vehicle can lead to performance issues or engine damage. Always double-check the part number to ensure compatibility with your car.
- **Forgetting the Gap Check:** Improperly gapped plugs can lead to poor performance. Take the time to measure and adjust them.

Replacing your spark plugs ensures smoother engine operation, improved fuel efficiency, and better overall performance. Plus, it's an empowering project that takes less than an hour and requires only a few tools. By learning to change your own spark plugs, you're not just saving money—you're taking control of your car's maintenance.

Replacing a Fuel Filter

Replacing your car's fuel filter is an important maintenance task that ensures your engine receives clean fuel, free from contaminants that could reduce performance or damage components. Thankfully, this is a task you can handle yourself with a little preparation and care.

Tools Needed:

- New fuel filter (ensure it's the correct type for your car)
- Wrenches (or a fuel line disconnect tool)
- Safety gloves and goggles
- Drain pan or container

Estimated Time: 30–45 minutes

Instructions:

- **Depressurize the Fuel System:** Fuel systems are pressurized, so you'll need to relieve that pressure before starting. Locate the fuel pump fuse or relay (check your owner's manual) and remove it. Start the engine and let it run until it stalls—this ensures any remaining pressure in the fuel line is released. Turn off the ignition and disconnect the battery for extra safety.

- **Locate the Fuel Filter:** Fuel filters are typically along the fuel line, either under the car (near the fuel tank) or in the engine bay. Your owner's manual will help you pinpoint its exact location.

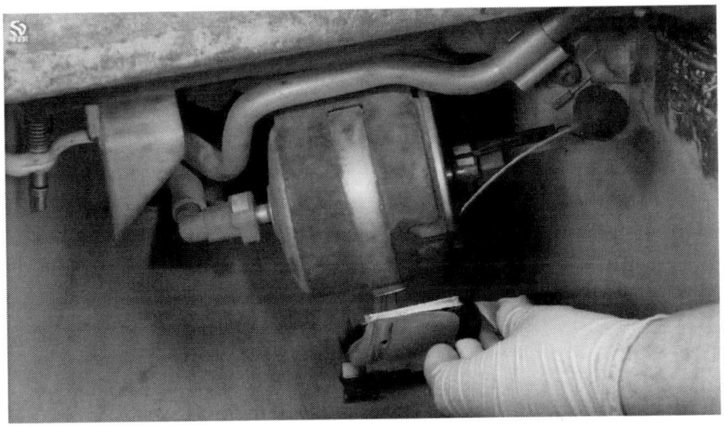

- **Remove the Old Filter:** Position a drain pan under the filter to catch any fuel that may spill. Use wrenches or a fuel line disconnect tool to loosen the clamps securing the fuel lines to the filter. Carefully detach the lines—expect a small amount of fuel to leak. Remove the old filter, noting its orientation, especially the direction of fuel flow.

- **Install the New Filter:** Take the new filter and ensure it's oriented correctly. Most filters have an arrow indicating the direction of fuel flow. Attach the filter to the fuel lines, tighten the clamps securely, and double-check all connections.

- **Test for Leaks:** Reinstall the fuel pump fuse or relay. Start the engine and let it idle while inspecting the filter connections for leaks. If you spot any fuel seepage, tighten the clamps slightly until the connections are secure.

Safety Precautions:

- **Work in a ventilated area**: Fuel fumes can be hazardous, so ensure good airflow.
- **Avoid open flames**: Keep lighters, cigarettes, or anything that might spark far away.
- **Wear protective gear**: Gloves and goggles will protect your skin and eyes from fuel contact.

Common Pitfalls and Troubleshooting Tips:

- **Reversing the filter orientation**: Installing the filter backward blocks fuel flow. Check the arrow on the filter for proper alignment.
- **Forgotten fuse or relay**: If the engine doesn't start after replacing the filter, confirm the fuel pump fuse or relay has been reinstalled.
- **Over-tightening clamps**: While tight connections are critical, excessive force can damage the filter or fuel lines.

Over time, your fuel filter traps dirt, rust, and debris, ensuring only clean fuel reaches your engine. Replacing it not only improves performance but also extends the life of critical engine components. Plus, it's a satisfying repair that you can confidently tackle yourself.

Flushing and Refilling Coolant

Replacing your car's coolant isn't just about keeping your engine at the right temperature—it also prevents corrosion and ensures the cooling system works efficiently. Thankfully, flushing and refilling your coolant is a task you can handle with a bit of patience and care.

Tools Needed:

- Coolant (refer to your manual for the correct type)
- Distilled water (for flushing)
- Funnel
- Drain pan
- Wrench or screwdriver (for drain plug or hose clamps)
- Gloves and safety goggles

Estimated Time: 45–60 minutes

Instructions:

- **Cool the Engine:** First things first—safety matters. Wait until your engine is completely cool before starting. Opening the radiator cap while the engine is hot can release pressurized coolant, causing burns. Patience is key here.
- **Drain the Old Coolant:** Place your drain pan underneath the radiator. Locate the radiator drain plug or the lower radiator hose, depending on your vehicle. Loosen the plug or hose clamp carefully, and let the old coolant flow into the pan. Once the system is empty, close the drain plug or reconnect the hose.

- **Flush the System:** Now, it's time to clear out any remaining debris or old coolant. Close the drain plug and fill the radiator with distilled water. Start the engine and let it run for 10–15 minutes to circulate the water through the system. Turn off the engine, let it cool again, and then drain the water just as you did with the coolant.

- **Refill with New Coolant:** Use a funnel to pour fresh coolant into the radiator. Check your manual for the correct coolant-to-water ratio if mixing is required. Most modern coolants come pre-mixed, so you may not need to add water. Don't forget to fill the overflow reservoir to the recommended level.

- **Bleed Air from the System:** Start the engine and run it with the radiator cap off. This allows any trapped air to escape, which prevents overheating later. Keep an eye on the coolant level, as it may drop while the air escapes. Add more coolant as needed until the level stabilizes. Once you're done, securely replace the radiator cap.

Safety Precautions:

- **Wear gloves and goggles**: Coolant is toxic and can irritate your skin and eyes.
- **Dispose of old coolant responsibly**: Don't pour it down the drain—take it to a recycling center or auto shop.
- **Work in a ventilated area**: Coolant fumes can be harmful in enclosed spaces.

Common Pitfalls and Troubleshooting Tips:

- **Trapped Air:** One of the most common issues is air remaining in the system, which can lead to overheating. Take your time during the bleeding process to ensure all air bubbles are out.
- **Wrong Coolant Type:** Using the incorrect type of coolant can damage your engine or cooling system. Double-check your owner's manual to confirm the right formula for your vehicle.
- **Leaky Connections:** After refilling, inspect all connections, including the drain plug and hoses, for leaks. Tighten any loose clamps if necessary, but avoid overtightening, which can cause damage.

Regularly flushing and refilling your coolant keeps your engine running efficiently and prevents costly damage from overheating or corrosion. Plus, this simple maintenance task gives you a chance to check the overall health of your cooling system.

Replacing Radiator Hoses

Radiator hoses are critical for circulating coolant between your radiator and engine. Over time, these hoses can crack, swell, or become brittle, leading to leaks and overheating. Replacing them when signs of wear appear is a straightforward task that can save you from expensive repairs down the road.

Tools Needed:

- New radiator hoses (upper and lower)
- Screwdriver or pliers
- Drain pan
- Coolant (refer to your manual for the correct type)
- Funnel
- Safety gloves and goggles

Estimated Time: 30–45 minutes

Instructions:

- **Drain the Coolant:** Start by ensuring the engine is completely cool to avoid burns. Place a drain pan under the radiator. Open the radiator drain plug or loosen the lower hose to let the coolant flow out. Be patient—this may take a few minutes. Once it's drained, secure the plug or hose.

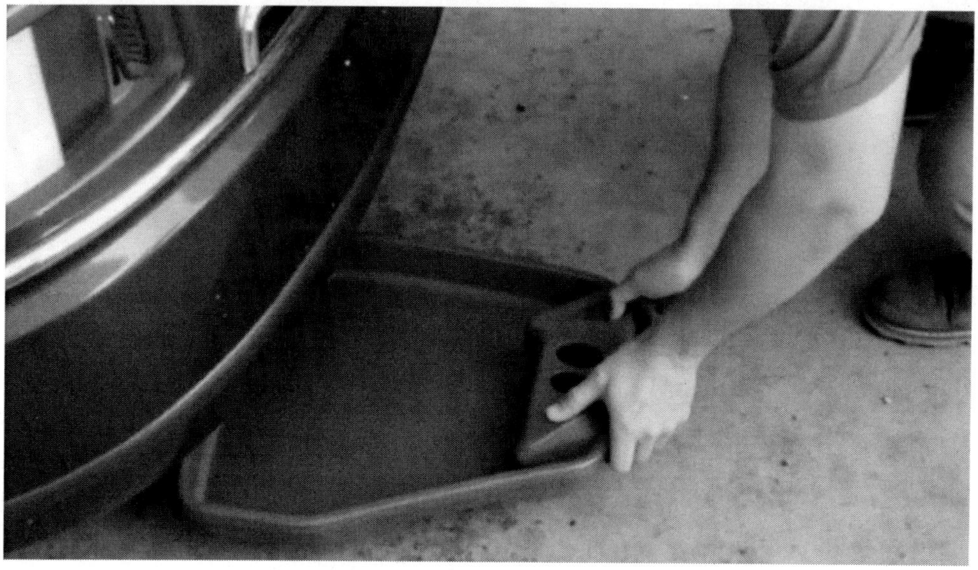

- **Remove the Old Hoses:** Locate the upper and lower radiator hoses. Use a screwdriver or pliers to loosen the clamps securing each hose. Carefully twist and pull the hoses off the radiator and engine fittings. Be prepared for a little leftover coolant to spill out. Check the fittings for any residue or corrosion, and clean them if necessary.

- **Install the New Hoses:** Take your new hoses and slide them onto the radiator and engine fittings. Ensure they are aligned properly and pushed all the way onto the fittings. Secure the hoses with the clamps, tightening them snugly. Avoid overtightening, as this can damage the hose or clamp.

- **Refill the Coolant:** Using a funnel, refill the radiator with fresh coolant. Don't forget to top off the overflow reservoir as well. If your coolant requires mixing, ensure the water-to-coolant ratio matches your vehicle's specifications.

- **Bleed Air from the System:** Start the engine and let it idle with the radiator cap off. This allows air bubbles to escape. Watch the coolant level and add more as needed. Once air stops escaping, replace the radiator cap securely.

- **Inspect for Leaks:** With the engine running, check around the hose connections for any leaks. Tighten clamps slightly if you notice drips. Turn off the engine and let it cool before performing a final inspection.

Safety Precautions:

- Wait for the engine to cool completely: Hot coolant can cause severe burns.
- Use gloves and goggles: Protect your hands and eyes from coolant and sharp edges.
- Dispose of old coolant responsibly: Coolant is toxic and must be taken to a recycling center or auto shop.

Common Pitfalls and Troubleshooting Tips:

- **Loose or Overtightened Clamps:** Make sure clamps are tight enough to prevent leaks but not so tight that they cut into the hose. A snug fit is ideal.
- **Not Checking for Leaks:** Skipping a leak check can lead to coolant loss and overheating. Always inspect your work before hitting the road.
- **Trapped Air:** Air bubbles in the system can cause uneven cooling. Spend extra time bleeding the system to avoid overheating issues.

Replacing your radiator hoses is more than just routine maintenance—it's a vital step in ensuring your car's cooling system works efficiently. Worn-out hoses can lead to sudden failures, leaving you stranded with an overheated engine. By taking the time to replace them, you're protecting your engine and ensuring peace of mind on the road.

Replacing the Thermostat

Replacing a thermostat is a straightforward repair that can resolve issues like overheating or inefficient heating in your car. This small component plays a big role in regulating your engine's temperature, so ensuring it's in good working order is essential.

Tools Needed:

- New thermostat and gasket (refer to your car's manual for compatibility)
- Socket wrench and sockets
- Screwdriver
- Drain pan
- Coolant (check your manual for the correct type)
- Gasket scraper or razor blade
- Funnel
- Safety gloves and goggles

Estimated Time: 45–60 minutes

Instructions:

- **Cool the Engine:** Before starting, ensure the engine is completely cool. Working on a hot engine can lead to burns and make handling coolant dangerous.
- **Drain the Coolant:** Place a drain pan under the radiator. Open the drain plug (typically at the bottom of the radiator) or loosen the lower radiator hose to let the coolant flow out. Let it drain completely to avoid spills when removing the thermostat housing.

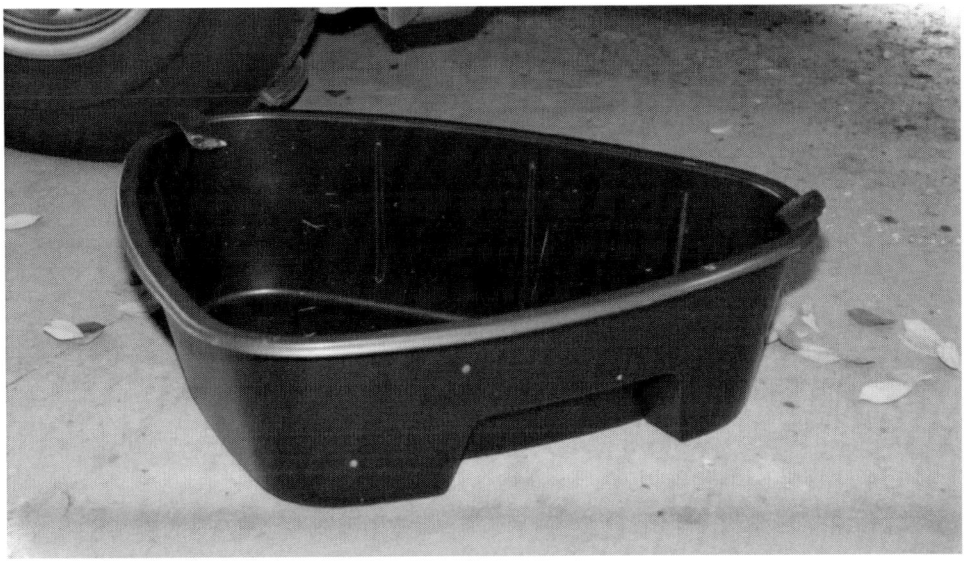

- **Locate the Thermostat:** Follow the upper radiator hose to its connection point with the engine. The thermostat is housed where the hose connects to the engine block.

- **Remove the Thermostat Housing:** Using a socket wrench, loosen the bolts securing the housing. Lift the housing carefully, as coolant might still be inside. Once removed, you'll see the thermostat sitting inside the engine.

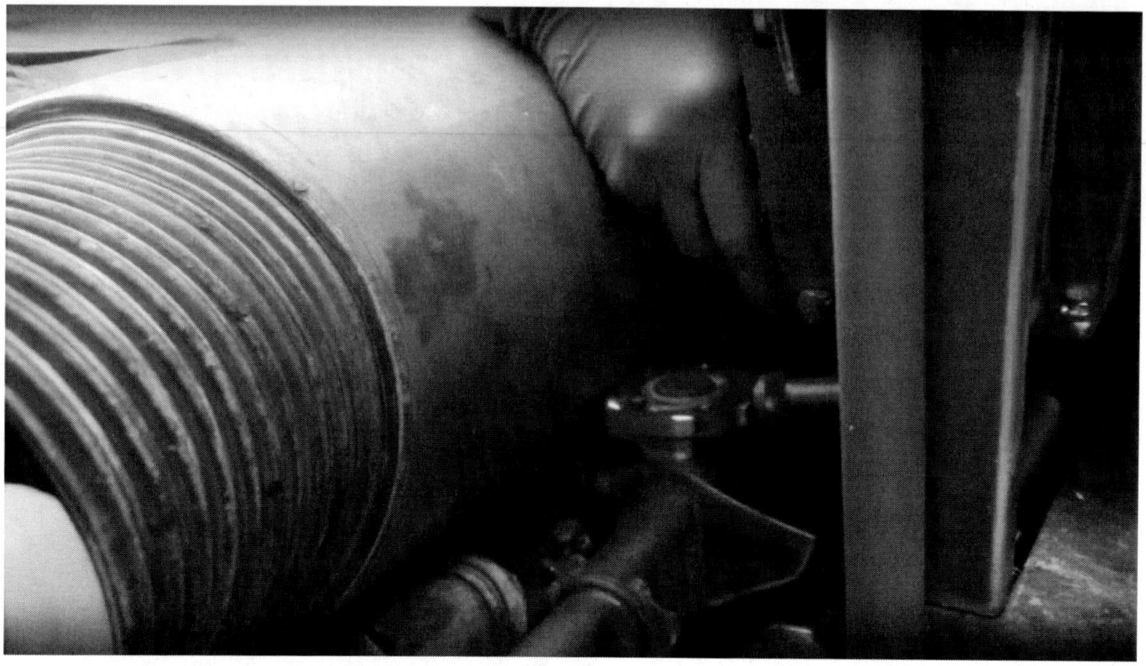

- **Remove the Old Thermostat:** Pull the old thermostat out and set it aside. Remove the gasket (if present) and use a gasket scraper or razor blade to clean any residue on the housing and engine mating surfaces. A clean surface ensures a proper seal.

- **Install the New Thermostat:** Insert the new thermostat in the same orientation as the old one, with the spring side facing the engine. Place the new gasket onto the housing or engine, aligning it properly to prevent leaks.

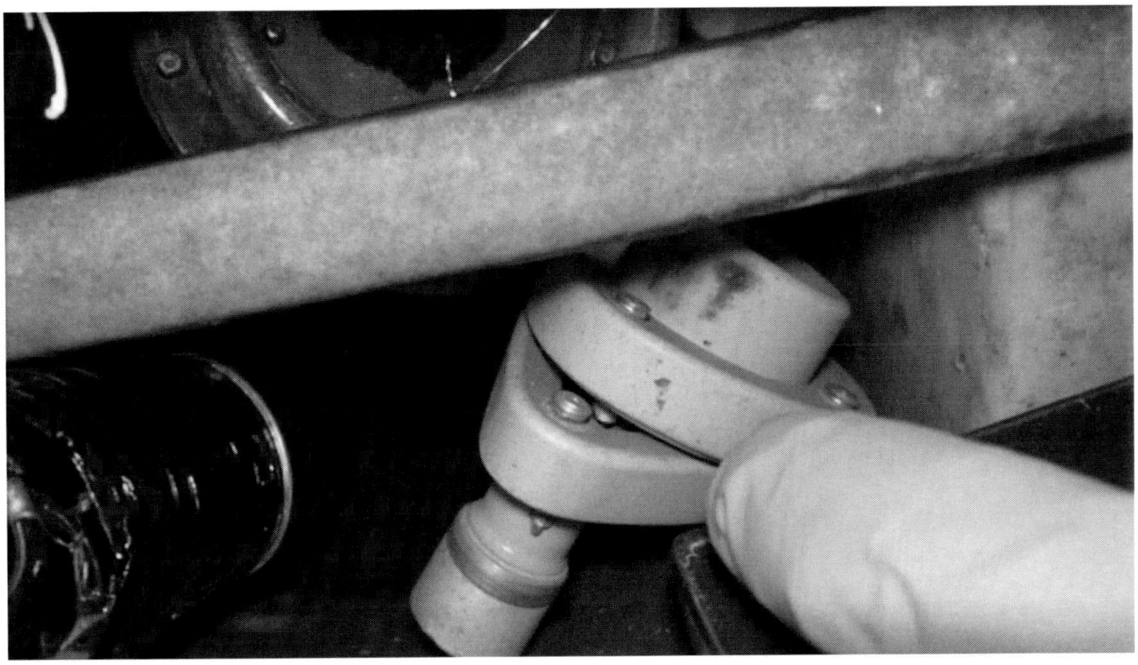

- **Reattach the Thermostat Housing:** Reposition the housing over the thermostat and secure it with bolts. Tighten them evenly, but don't overtighten—this could crack the housing.

- **Refill the Coolant:** Using a funnel, add fresh coolant to the radiator. Top off the overflow reservoir to the recommended level.

- **Bleed Air from the System:** Start the engine and let it idle with the radiator cap off. As the engine warms up, trapped air will escape as bubbles in the coolant. Once the bubbles subside, add more coolant if needed and securely replace the radiator cap.

- **Check for Leaks:** Inspect the thermostat housing and nearby hoses for any signs of leaks. If everything looks secure, take the car for a short drive and monitor the temperature gauge to ensure the system is functioning properly.

Safety Precautions:

- **Wait for the engine to cool:** Never open the radiator cap or work near coolant while the engine is warm.
- **Wear gloves and goggles:** Coolant is toxic, so avoid skin contact and protect your eyes.
- **Dispose of coolant responsibly:** Used coolant must be taken to a recycling center or auto shop.

Common Pitfalls and Troubleshooting Tips:

- **Incorrect Orientation:** Make sure the new thermostat is installed in the correct orientation, typically with the spring side facing the engine. Refer to your manual if you're unsure.
- **Leaks:** Residual gasket material can cause leaks. Take your time to clean the mating surfaces thoroughly before installing the new gasket.
- **Air Pockets:** Failing to bleed air from the cooling system can lead to overheating. After finishing the repair, keep an eye on the temperature gauge during your test drive.

A faulty thermostat can lead to overheating or inefficient cooling, which may result in engine damage if ignored. By replacing the thermostat yourself, you're not only saving money but also ensuring your cooling system operates smoothly. Plus, with a little patience and preparation, this repair can be a confidence booster for tackling more advanced projects.

Adjusting the Handbrake

A properly adjusted handbrake is essential for keeping your car secure when parked, especially on inclines. Over time, handbrake cables can stretch, or the braking system can wear, leaving your handbrake loose or ineffective. The good news? Adjusting it is a relatively simple task you can tackle at home.

Tools Needed:

- Jack and jack stands
- Socket wrench or spanner set
- Flathead screwdriver
- Gloves and safety goggles

Estimated Time: 30–45 minutes

Instructions:

- **Secure the Car:** Park your car on level ground. Engage the foot brake, set the gear to "Park" (or put it in gear for manual transmissions), and place wheel chocks under the tires for added safety.

- **Lift the Car:** Using a jack, lift the rear of the car so you can access the rear wheels or the undercarriage. Secure the vehicle with jack stands to ensure it's stable and safe to work underneath.

- **Locate the Handbrake Adjuster:** The location of the adjuster depends on your car model:
  - **Inside the Car**: Many vehicles have an adjustment nut near the handbrake lever. To access it, remove the center console or the cover surrounding the lever.
  - **Under the Car**: In some models, the handbrake adjuster is located underneath the car, near the rear wheels or the cable splitter.

Check your owner's manual if you're unsure.

- **Tighten the Cable:** Turn the adjustment nut clockwise using a socket wrench or spanner. Tighten in small increments, testing the handbrake lever's tension after each adjustment. Aim for the handbrake to engage fully after 3–5 clicks.

- **Check Each Wheel:** With the handbrake engaged, try to spin the rear wheels by hand. They should not move. If the wheels still spin, tighten the cable slightly more. Once tightened, disengage the handbrake and spin the wheels again to ensure they move freely.
- **Lower the Car:** After testing and ensuring the adjustment is correct, carefully lower the car off the jack stands. Double-check the handbrake's functionality on flat ground and a slight incline before hitting the road.

Safety Precautions:

- Always secure the car on jack stands before working underneath it.
- Wear gloves to protect your hands from pinching hazards near the brake components.
- Do not overtighten the handbrake cable, as this can lead to premature brake wear.

Common Pitfalls and Troubleshooting Tips:

- **Overtightening:** If the handbrake is too tight, it can cause the brakes to drag while driving, which leads to unnecessary wear and reduced fuel efficiency. To avoid this, always test the rear wheels to ensure they spin freely when the handbrake is disengaged.
- **Uneven Adjustment:** Ensure both rear wheels are evenly adjusted. Unequal tension can result in one wheel locking up before the other, reducing braking performance and safety.

- **Persistent Loose Handbrake:** If tightening the cable doesn't solve the issue, the problem may lie elsewhere. Worn brake pads or shoes can reduce the handbrake's effectiveness and may need replacement.

Adjusting your handbrake improves your car's safety and ensures reliable performance, especially when parking on slopes. With just a few tools and some attention to detail, you'll restore your handbrake's functionality and gain confidence in your DIY car repair skills.

ADVANCED PROJECTS: REPAIRS TO CHALLENGE YOUR SKILLS

These tasks require more tools, time, and attention to detail. If you've mastered the intermediate projects, you're ready to tackle these.

Replacing Shocks and Struts

If your car feels bouncy, leans too much during turns, or you're noticing uneven tire wear, it might be time to replace the shocks or struts. These components absorb impacts and stabilize your car, providing a smoother and safer ride. While this project takes time and some patience, the rewards are well worth the effort.

Tools Needed:

- Jack and jack stands
- Socket set and ratchet
- Wrenches
- Spring compressor (for struts)
- Torque wrench
- Safety gloves and goggles

Estimated Time: 2–4 hours

Instructions:

- **Lift and Secure the Car:** Start by parking on a flat, stable surface and engaging the parking brake. Use a jack to lift the car, then place jack stands securely underneath. Never rely solely on the jack—it's not worth the risk.
- **Remove the Wheels:** Use a socket wrench to remove the lug nuts, then take off the wheels to expose the shocks or struts. Keep the lug nuts in a safe spot—you'll need them later.
- **Detach the Old Shock or Strut:**

- **For Shocks**: Locate the bolts at the top and bottom of the shock. Remove them using your socket set or wrenches.

- **For Struts**: Remove the bolts securing the strut to the steering knuckle. Then, open the hood and locate the bolts securing the top of the strut in the engine bay. Remove these bolts carefully, supporting the strut as it detaches.

- **Compress the Spring (Struts Only):** If you're replacing a strut, use a spring compressor to release the tension on the coil spring. This step is critical for safety—coil springs store a lot of energy and can be dangerous if not handled properly. Once the spring is compressed, remove it from the old strut assembly.

- **Install the New Shock or Strut:** For both shocks and struts, position the new component in place and secure it with the bolts you removed earlier. Tighten the bolts to the manufacturer's torque specifications using a torque wrench. This ensures the parts are properly secured without overtightening.

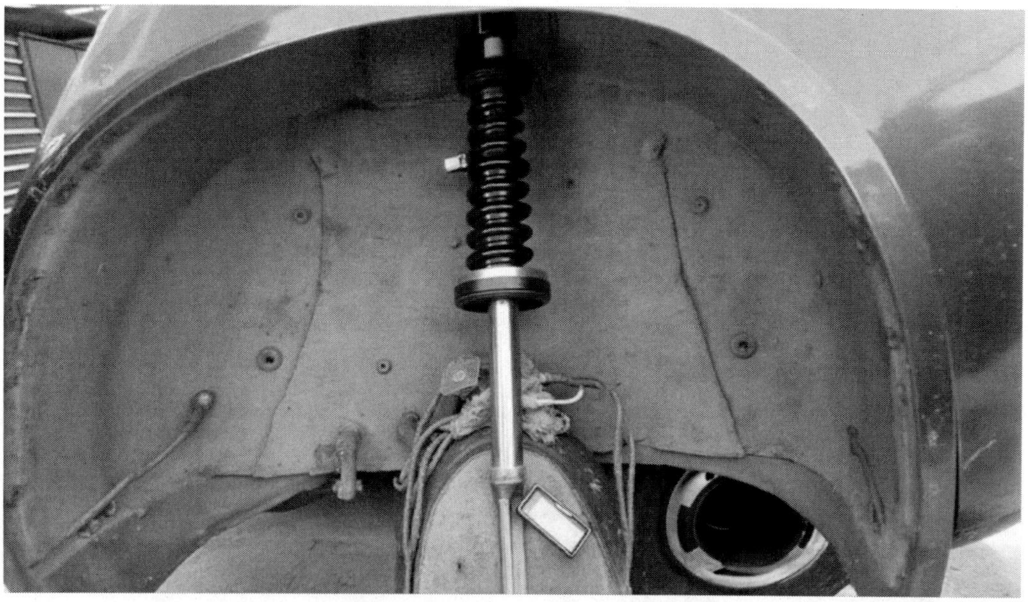

- **Reassemble and Test:** Reinstall the wheels, tighten the lug nuts, and lower the car from the jack stands. Take your car for a cautious test drive to ensure the ride feels smooth and stable.

Safety Precautions:

- **Use jack stands**: Never work under a car supported only by a jack.
- **Handle springs with care**: If working on struts, always use a spring compressor. Coil springs can cause serious injury if released suddenly.
- **Wear protective gear**: Gloves and goggles are a must to protect your hands and eyes from debris or slipping tools.

Common Pitfalls and Troubleshooting Tips:

- **Skipping Torque Specs:** It's tempting to estimate tightness, but over- or under-tightening bolts can lead to poor performance or damage. Always use a torque wrench to meet the manufacturer's specifications.
- **Not Replacing in Pairs:** Always replace shocks or struts on both sides of the car, even if only one seems worn. This ensures even handling and prevents uneven wear on other suspension components.
- **Forgetting the Alignment:** Replacing struts can affect your car's alignment. After completing the job, it's a good idea to have your alignment checked to ensure proper handling and even tire wear.

Replacing your shocks and struts not only improves ride comfort but also enhances your car's safety and stability. You'll notice better handling, especially during turns or on bumpy roads. By tackling this project yourself, you'll save on labor costs while gaining confidence in your repair skills.

Replacing an Alternator

Your car's alternator plays a critical role—it keeps your battery charged and powers the electrical systems while the engine is running. If your headlights are dimming, your battery keeps dying, or you're noticing electrical issues, it might be time for a replacement. While the task may seem daunting, it's a manageable project with the right tools and a bit of patience.

Tools Needed:

- Socket set and ratchet
- Wrenches
- Belt tensioner tool
- Screwdrivers
- Safety gloves and goggles

Estimated Time: 1–2 hours

Instructions:

- **Disconnect the Battery:** Always start by disconnecting the negative battery terminal. This is a safety must to avoid electrical shorts or sparks while working on the alternator.
- **Locate the Alternator:** The alternator is usually found at the front of the engine, connected to the serpentine belt. Refer to your car's manual if you're having trouble spotting it.

- **Loosen the Belt:** Using a belt tensioner tool, release the tension on the serpentine belt. Once loose, slip the belt off the alternator pulley. Take note of the belt's routing or snap a photo—it will need to go back exactly the same way.
- **Disconnect the Wires:** Unbolt all wires connected to the alternator. This typically includes:
    - The main power cable (held by a nut).
    - One or more smaller connectors (these may require a screwdriver to remove).

Keep track of which wire goes where—label them if needed.

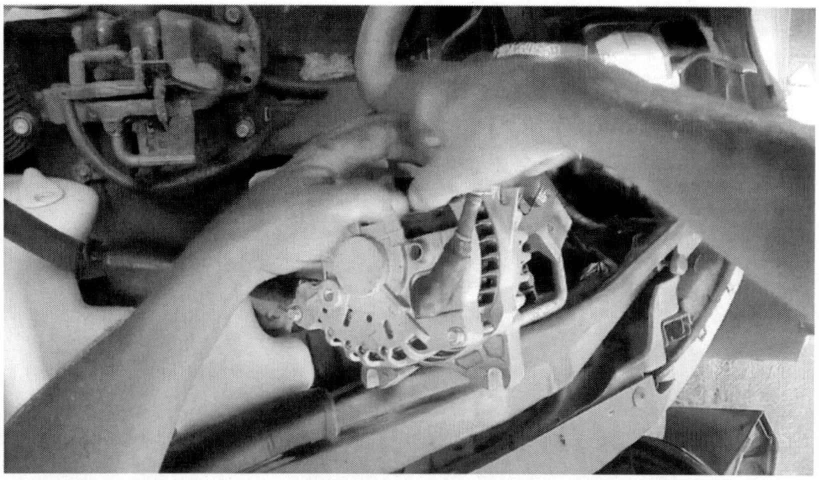

- **Remove the Old Alternator:** Unscrew the bolts holding the alternator in place. Once all bolts are removed, carefully lift the alternator out of its bracket.

- **Install the New Alternator:** Position the new alternator in the bracket and secure it with the bolts. Tighten them snugly but don't overtighten—refer to your manual for torque specifications if available.

- **Reinstall the Belt:** Slip the serpentine belt back onto the alternator pulley, ensuring it's aligned with all other pulleys. Use the belt tensioner tool to tighten the belt to the correct tension.
- **Reconnect the Battery and Test:** Reattach the negative battery terminal. Start the car and let it idle. Check for proper alternator function—your dashboard battery light should go off, and accessories like lights and radio should operate without issue.

Safety Precautions:

- Never work on the alternator while the engine is running—this can cause injury or damage.
- Wear gloves and goggles to protect against accidental cuts or electrical sparks.
- Ensure the serpentine belt is properly tensioned to avoid slippage.

Common Pitfalls and Troubleshooting Tips:

- **Loose or Poorly Connected Wires:** Double-check that all wires are securely connected and correctly positioned. A loose wire can prevent the alternator from charging the battery.
- **Misaligned Belt:** Ensure the serpentine belt is sitting properly on all pulleys. A misaligned belt can cause wear, noise, or slippage.
- **Still Having Issues?** If the new alternator doesn't resolve your charging problems, test the battery or inspect the voltage regulator (often built into the alternator).

Replacing your alternator can save you hundreds of dollars in labor costs and ensure your car's electrical systems function smoothly. It's a moderately challenging task, but with clear instructions and the right mindset, you'll feel confident tackling it.

Fixing a Faulty Starter

A faulty starter can leave you stranded, unable to start your car even with a fully charged battery. Replacing it might sound intimidating, but with the right tools and a systematic approach, you can handle this repair yourself and save a good chunk of money.

Tools Needed:

- Jack and jack stands
- Socket set and ratchet
- Wrenches
- Safety gloves and goggles

Estimated Time: 1–2 hours

Instructions:

- **Disconnect the Battery:** Begin by disconnecting the negative battery terminal. This ensures you avoid accidental sparks or electrical shorts while working on the starter.
- **Locate the Starter:** The starter is typically found near the bottom of the engine, close to the transmission. If you're having trouble, consult your car's manual for the exact location.

- **Remove the Wiring:** Carefully disconnect all wiring attached to the starter. This usually includes:
    - A thick power cable connected to the battery.
    - One or more smaller wires for the solenoid.
    - Take note of where each wire connects so you can reinstall them correctly.

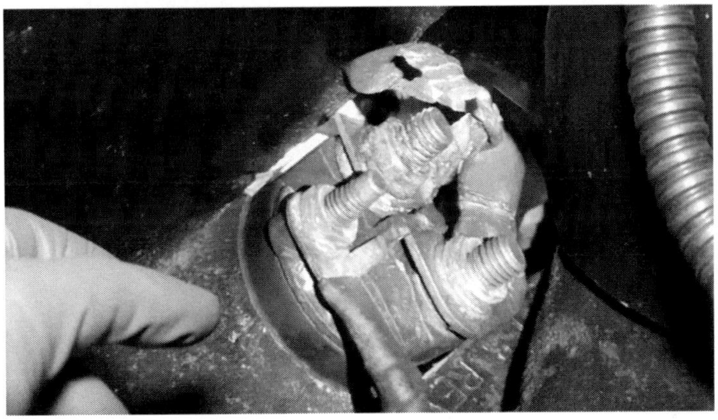

- **Unbolt the Starter:** Use a socket wrench to remove the bolts securing the starter to the engine or transmission housing. Once the bolts are out, carefully slide the starter out of its position.

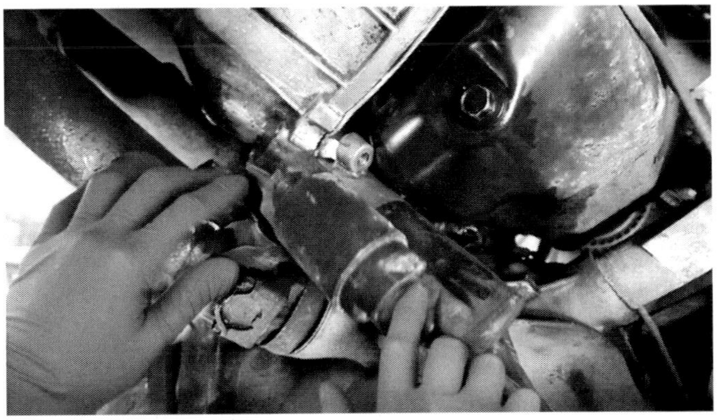

- **Install the New Starter:** Position the new starter in the same spot as the old one. Align it properly and secure it with the bolts you removed earlier. Tighten them snugly, but don't overtighten—refer to your car's manual for torque specifications if available.

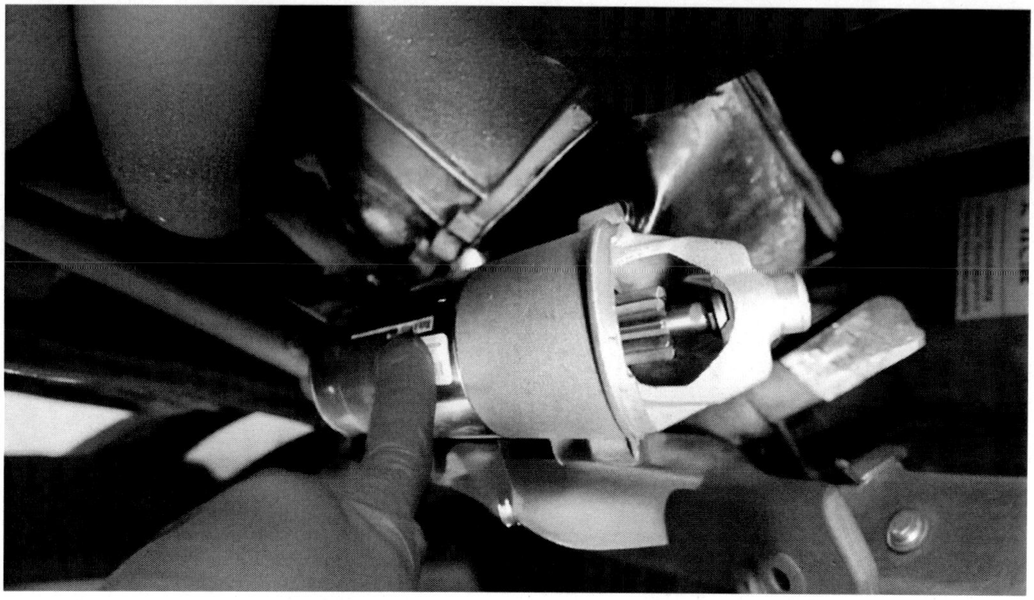

- **Reconnect the Wiring:** Reattach the power cable and smaller wires to the correct terminals on the new starter. Double-check that each connection is secure before moving on.

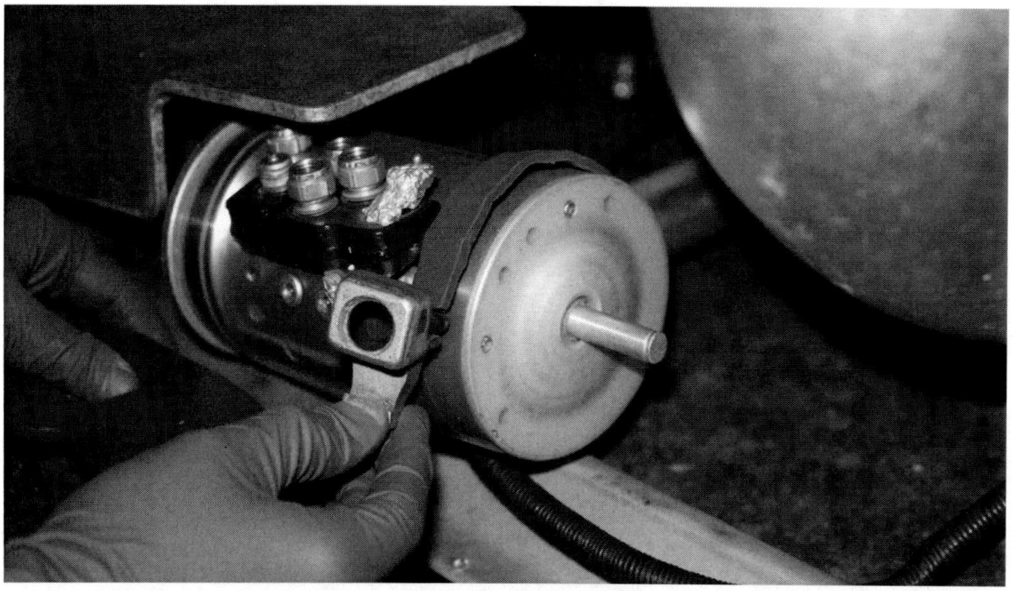

- **Reconnect the Battery and Test:** Reconnect the negative battery terminal, then start your car to test the new starter. If everything is working properly, the engine should crank smoothly and start without hesitation.

Safety Precautions:

- Always secure the car on jack stands before working underneath it. Never rely solely on the jack for support.
- Avoid touching any hot components if the car was running recently. Allow the engine to cool down before starting the repair.
- Wear gloves and goggles to protect against accidental scrapes or electrical hazards.

Common Pitfalls and Troubleshooting Tips:

- **Loose or Poor Connections:** Ensure all wires are tightly secured to the starter. Loose connections can prevent the starter from engaging or cause intermittent issues.
- **Wrong Starter Fit:** Always confirm that the replacement starter matches your car's make, model, and engine specifications. An incorrect fit may not align properly or function as needed.
- **Still Won't Start?** If the new starter doesn't fix the issue, check the battery, ignition switch, and relays for potential problems.

Replacing a starter might sound like a job for a professional, but it's a straightforward task that can save you hundreds of dollars in labor costs. Plus, tackling this project on your own gives you a deeper understanding of your car's electrical system and boosts your confidence for future repairs.

Inspecting and Replacing CV Joints and Boots

Your car's CV joints and boots are critical for transferring power to the wheels while accommodating the suspension's movement. Over time, these components can wear out or develop cracks, leading to grease leaks and eventual failure. Replacing a damaged CV joint or boot early can save you from more costly repairs later.

Tools Needed:

- Jack and jack stands
- Socket set and ratchet
- Pliers
- CV joint grease
- Replacement boots or CV joint kit
- Safety gloves and goggles

Estimated Time: 2–4 hours

Instructions:

- **Lift the Car and Remove the Wheel:** Park your car on a level surface and engage the parking brake. Use a jack to lift the vehicle and secure it with jack stands. Remove the wheel to access the CV joint and boot.
- **Inspect the Boot for Damage:** Thoroughly examine the CV boot for cracks, tears, or signs of leaking grease. A damaged boot often indicates that dirt and debris may have entered the joint, which can accelerate wear.

- **Remove the Axle (for CV Joint Replacement):** If you need to replace the entire CV joint, you'll first need to remove the axle:
  - Unbolt the axle from the transmission and wheel hub.
  - Carefully slide the axle out. Be gentle to avoid damaging the seals at the transmission end.

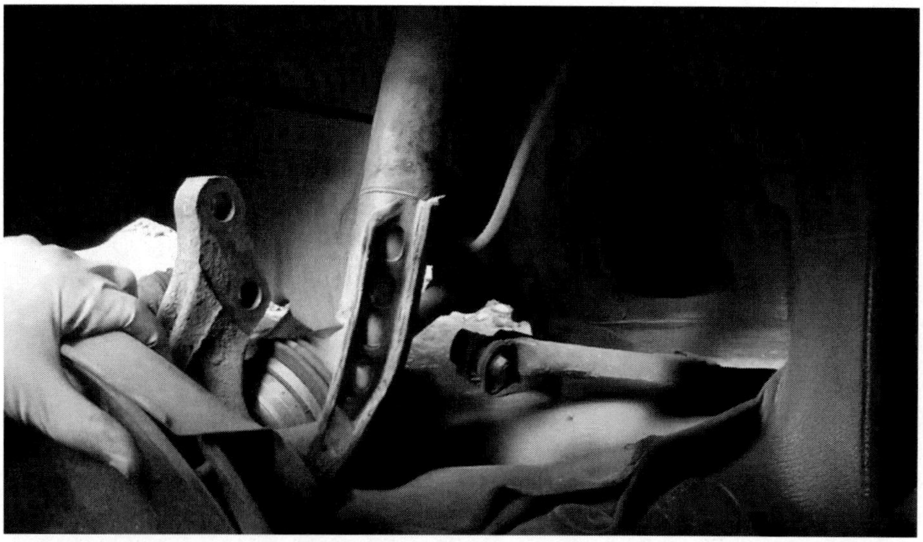

- **Replace the CV Boot:** If only the boot is damaged, clean the CV joint thoroughly and inspect it for wear. If the joint itself is intact:
  - Apply fresh CV joint grease generously.
  - Slide the replacement boot over the joint and secure it with the provided clamps. Ensure the clamps are tight to prevent grease leaks.
- **Replace the CV Joint (if Needed):** If the CV joint is damaged or has significant wear:
  - Follow the instructions provided with your CV joint replacement kit.
  - Clean all components, apply fresh grease, and assemble the new joint as directed.

- **Reinstall the Axle and Wheel:** Slide the axle back into the transmission and wheel hub, ensuring it is properly seated. Bolt it securely in place and reattach the wheel. Lower the car off the jack stands.

Safety Precautions:

- Always wear gloves when handling greasy components to protect your hands.
- Ensure the car is fully stable on jack stands before removing any parts.
- Use caution when sliding the axle out to avoid damaging seals, which could lead to transmission leaks.

Common Pitfalls and Troubleshooting Tips:

- **Skipping the Grease:** Neglecting to clean and reapply grease can lead to premature CV joint failure. Fresh grease is essential for smooth operation.
- **Damaging the Seals:** Be careful when removing and reinstalling the axle. Damaging the seals at the transmission end can result in fluid leaks, which can cause further issues.
- **Clamps Too Loose or Too Tight:** When securing the new boot, ensure the clamps are tight enough to prevent grease from leaking but not so tight that they cut into the rubber.
- **Ignoring Early Signs of Wear:** Clicking sounds during turns or visible grease leaks are signs of CV joint issues. Address these early to avoid more extensive repairs.

Replacing a CV boot or joint yourself might seem challenging, but it's a highly rewarding project that can save you significant labor costs. Plus, maintaining these components ensures smoother handling and prevents damage to other drivetrain parts.

Changing Automatic Transmission Fluid

Keeping your automatic transmission fluid clean and at the correct level is essential for smooth gear shifts and a longer transmission lifespan. While it might sound intimidating, this is a manageable project for any DIYer willing to follow some straightforward steps.

Tools Needed:

- Jack and jack stands
- Socket set and ratchet
- Drain pan
- Funnel
- New transmission fluid (consult your owner's manual for the correct type)
- Torque wrench

Estimated Time: 1–2 hours

Instructions:

- **Warm Up the Car:** Start your engine and let it run for a few minutes. Warming up the transmission fluid makes it less viscous, so it drains more easily. However, allow it to cool slightly to avoid burns when you're ready to begin.
- **Lift and Secure the Car:** Park on a flat surface, engage the parking brake, and place wheel chocks under the tires for added safety. Use a jack to lift the car and secure it with jack stands. Ensure the car is stable before working underneath.

- **Drain the Old Transmission Fluid:** Locate the transmission pan under your vehicle. Place a drain pan beneath it, then use a socket wrench to remove the drain plug. Let the old fluid flow out completely. Be prepared for a steady stream of fluid, and keep the pan positioned to catch every drop.

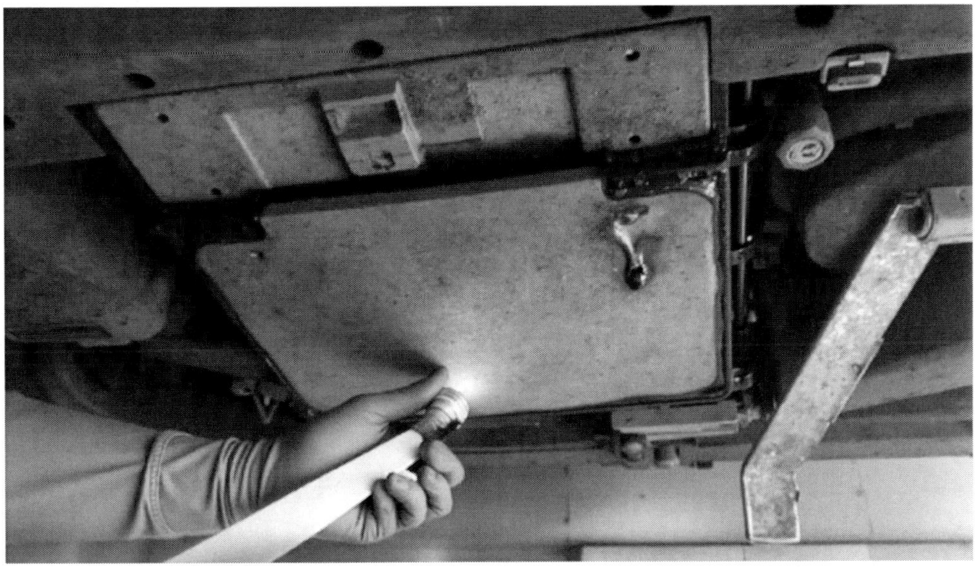

- **Replace the Filter (If Applicable):** Not all cars have easily replaceable transmission filters, and some vehicles, particularly those with Continuously Variable Transmissions (CVTs) or sealed transmissions, may require specialized service for filter replacement. If your vehicle allows for a straightforward filter change:
    - Remove the bolts securing the transmission pan. Be cautious, as some fluid may still be inside.
    - Once the pan is off, carefully remove the old filter and replace it with a new one. Secure the pan back in place, tightening the bolts to the manufacturer's torque specifications.

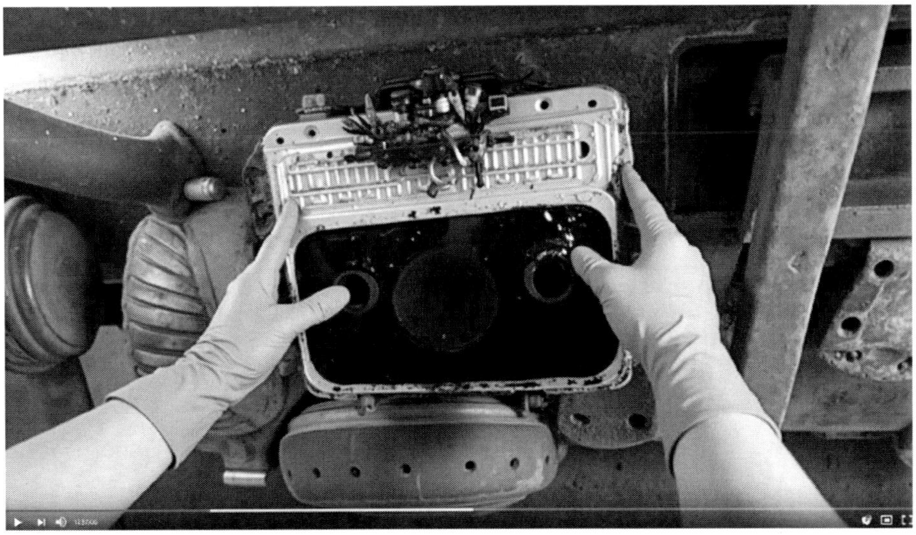

For vehicles with sealed systems or CVTs, consult your owner's manual or have the service performed by a professional to ensure proper handling and avoid damage.

- **Refill the Transmission Fluid:** After ensuring the drain plug is securely tightened, lower the car and locate the transmission dipstick tube. Insert a funnel and slowly pour in the new fluid. Start with the amount specified in your manual, but leave room to adjust.
- **Check the Fluid Level:** Start the engine and let it idle. With the engine running, shift through all the gears to circulate the new fluid. Return the car to park or neutral, then check the dipstick while the engine is still running. Add more fluid as needed, but avoid overfilling.

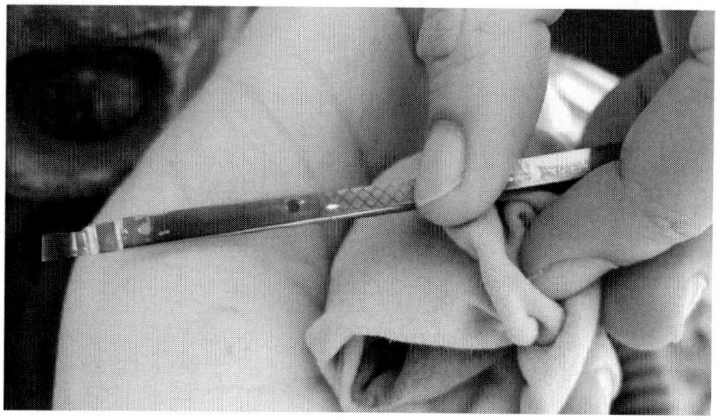

Safety Precautions:

- **Avoid burns:** Transmission fluid can get extremely hot. Let the engine cool slightly before starting.
- **Secure the vehicle:** Always use jack stands, not just a jack, for stability while working underneath.
- **Dispose of fluid properly:** Used transmission fluid is toxic. Take it to a recycling center or auto parts store for safe disposal.

Common Pitfalls and Troubleshooting Tips:

- **Overfilling:** Adding too much transmission fluid can cause it to foam, leading to shifting problems and potential damage. Always add fluid gradually and check the dipstick regularly.
- **Leaking Drain Plug:** After reinstalling the drain plug, double-check that it's tightened securely. A loose plug can lead to leaks, wasting your fresh fluid.
- **Skipping the Filter Replacement:** If your car has a replaceable filter and it's ignored, dirt and debris can remain in the system, negating the benefits of the fluid change.
- **Using the Wrong Fluid:** Not all transmission fluids are the same, and using an incompatible fluid can damage the transmission. Always refer to your owner's manual to ensure you're using the correct type for your vehicle. When in doubt, consult your manual or a trusted auto parts supplier to avoid costly mistakes.

Changing your automatic transmission fluid ensures smoother shifts, prevents overheating, and extends the life of your transmission. While it may seem like a job best left to a mechanic, this task is highly doable with a bit of patience and preparation. By taking it on yourself, you'll save money and gain valuable confidence in maintaining your vehicle.

Restoring the Cooling System with a Flush

Keeping your cooling system in top shape is essential to prevent engine overheating and costly repairs. Flushing the cooling system removes old, contaminated coolant and replaces it with fresh fluid, ensuring your car runs efficiently and stays cool.

Tools Needed:

- Jack and jack stands
- Drain pan
- Garden hose or distilled water
- Funnel
- New coolant (check your owner's manual for the correct type)
- Safety gloves and goggles

Estimated Time: 1–2 hours

Instructions:

- **Cool Down the Engine:** Never work on a hot engine—it's a serious burn hazard. Allow your engine to cool completely before starting.
- **Lift the Car:** Use a jack to raise the front of your car and secure it with jack stands. This gives you easier access to the radiator drain plug or lower radiator hose.
- **Drain the Old Coolant:** Place a drain pan beneath the radiator. Locate and open the radiator drain plug or remove the lower radiator hose. Let all the old coolant flow out. Once it's drained, replace the plug or hose securely.

- **Flush the System:** With the drain plug closed, fill the radiator with distilled water or connect a garden hose to run clean water through the system. Start the engine and turn the heater on full blast. Let it run for a few minutes to circulate the water, then turn the engine off and drain the water. Repeat this process until the water draining from the radiator runs clear.

- **Refill with Coolant:** Using a funnel, pour in fresh coolant. Most vehicles require a 50/50 mix of coolant and water but double-check your manual for specific instructions. Fill both the radiator and the overflow reservoir to the recommended levels.

- **Bleed the System:** Start the engine with the radiator cap off and allow it to idle. Turn the heater on to its highest setting. This helps air bubbles escape from the cooling system. Watch for bubbles in the radiator as the engine runs. Once the bubbles stop, top off the coolant and replace the cap.
- **Check for Leaks:** Inspect the radiator, hoses, and connections to ensure there are no leaks. A leak-free system is crucial to maintain proper pressure and cooling efficiency.

Safety Precautions:

- **Handle coolant carefully:** Coolant is toxic to humans and animals, so keep it out of reach of children and pets. Clean up spills immediately.
- **Dispose of coolant responsibly:** Take old coolant to a recycling center or auto parts store for proper disposal.
- **Wear gloves and goggles:** Protect your skin and eyes from contact with coolant, as it can be irritating.

Common Pitfalls and Troubleshooting Tips:

- **Skipping the Bleed Step:** Failing to bleed air from the system can leave air pockets, which may lead to overheating. Take your time to ensure all bubbles are gone.
- **Using the Wrong Coolant:** Not all coolants are the same. Mixing different types or using the wrong kind can damage your cooling system. Always use the type specified in your owner's manual.
- **Neglecting Hoses:** While flushing the system, inspect the radiator hoses for cracks or wear. Replace them if necessary to avoid future leaks.
- Flushing and refilling your cooling system is a straightforward task that delivers big benefits. It removes harmful deposits, prevents overheating, and extends the life of your radiator and engine. Plus, taking on this project yourself saves you money and builds your confidence for other car repairs.

Diagnosing and Replacing Fuses

A blown fuse can cause frustration, but fixing it is one of the simplest car repairs you can tackle. Fuses protect your car's electrical systems, acting as safety gates that "blow" if too much current flows through. Whether your radio suddenly cuts out, your headlights won't turn on, or your power windows stop working, a faulty fuse could be the culprit.

Tools Needed:

- Fuse puller (or needle-nose pliers)
- Multimeter (optional but useful for troubleshooting)
- Replacement fuses (matched to your car's specifications)

Estimated Time: 15–30 minutes

Instructions:

- **Locate the Fuse Box:** Start by finding your car's fuse boxes. Most vehicles have one under the hood and another inside the cabin, often near the driver's side dashboard or footwell. Check your owner's manual for precise locations.
- **Check the Diagram:** Each fuse box cover typically has a diagram or chart identifying which fuse corresponds to each system (e.g., lights, radio, or wipers). If the chart isn't visible, refer to your owner's manual. Locate the fuse tied to the issue you're troubleshooting.
- **Inspect the Fuse:** Using a fuse puller or needle-nose pliers, gently remove the fuse you suspect is blown. Examine the metal strip inside the fuse—it should form a continuous connection. If the strip is broken or the fuse looks burned, it needs replacing.
- **Replace with the Correct Fuse:** Match the blown fuse's amperage (indicated by a number on the fuse, such as 10A or 15A). Using a higher-rated fuse might seem like a quick fix, but it can lead to overheating or even an electrical fire. Insert the new fuse securely into the slot.
- **Test the System:** Turn the ignition back on and check the affected system. If your headlights now shine brightly, the radio plays or the power windows move smoothly, you've resolved the issue.

Safety Precautions:

- Turn off the ignition before working on the fuse box to prevent electrical shorts.
- Match fuse ratings precisely to avoid overloading the circuit.
- Work in a well-lit area so you can see small details clearly.

Common Pitfalls and Troubleshooting Tips:

- **Overlooking the Root Cause:** If a fuse repeatedly blows after replacement, the issue likely stems from a deeper problem, such as a short circuit, faulty wiring, or a malfunctioning component. Use a multimeter to test for abnormal currents or consult a professional if necessary.
- **Using the Wrong Fuse:** Inserting a fuse with the incorrect amperage can cause significant damage. Always double-check the fuse rating before replacing it. Keep extra fuses of various ratings in your car for emergencies.
- **Misidentifying the Problem:** Sometimes, the issue isn't the fuse itself but the system it protects. For example, a dead radio might be due to a loose connection rather than a blown fuse. If the fuse is intact but the problem persists, further investigation is needed.

Replacing a fuse is quick, inexpensive, and empowering. Learning to diagnose fuse issues not only saves you a trip to the mechanic but also builds your confidence to tackle other electrical problems. Keeping a stash of spare fuses and knowing how to replace them ensures you're ready to fix common issues on the go.

Diagnosing and Fixing Electrical Short Circuits

Electrical short circuits can be a frustrating challenge, but identifying and fixing them is well within your reach. Shorts happen when electricity flows where it shouldn't, usually due to exposed wires, faulty components, or improper connections. Tackling this problem requires patience and some basic tools, but the payoff is a well-functioning electrical system—and a better understanding of your car.

Tools Needed:

- Multimeter
- Fuse puller
- Electrical tape
- Replacement fuses
- Wire strippers

Estimated Time: 1–2 hours

Instructions:

- **Identify the Affected Circuit:** The first step is pinpointing the problem. Note which system isn't working properly (e.g., headlights, power windows, or radio). Open the fuse box and look for blown fuses. A blown fuse often has a melted or broken metal strip inside.

- **Test for Power Drain:** If the problem isn't immediately obvious, check for a parasitic drain. Disconnect the battery's negative terminal and set your multimeter to measure amperage. Connect the multimeter between the negative terminal and the disconnected cable. If the reading shows a higher-than-normal current draw, you likely have a short circuit.

- **Locate the Fault:** Next, you'll need to trace the issue to a specific circuit. With the multimeter still connected, pull fuses one at a time while monitoring the reading. When the amperage drops significantly, you've identified the circuit causing the problem.

- **Inspect Wires and Connectors:** Now it's time to look closely at the wires and connectors in the problematic circuit. Check for frayed insulation, melted plastic, exposed wires, or loose connections. Use electrical tape to patch small issues, or replace damaged wires with fresh sections. For more secure fixes, splice and solder wires before covering them with heat-shrink tubing.
- **Replace the Fuse and Test:** Once you've addressed the issue, replace the fuse with one of the correct amperage. Turn on the car and test the affected system. If everything works without blowing the fuse, congratulations—you've fixed the short circuit!

Safety Precautions:

- Always disconnect the battery before working on electrical systems to prevent accidental shocks or damage.
- Be careful when stripping wires or repairing connections to avoid cutting or pinching nearby wires.
- Work in a well-lit area to ensure you can see small details clearly.

Common Pitfalls and Troubleshooting Tips:

- **Skipping a Thorough Inspection:** Electrical shorts can be sneaky. Don't assume the first damaged wire you find is the only problem. Trace the circuit carefully to ensure all issues are addressed.
- **Using Poor Repairs:** Temporary fixes, like wrapping damaged wires in electrical tape, are fine for quick solutions but may not hold up over time. For a more permanent repair, splice wires securely and cover them with heat-shrink tubing.
- **Replacing the Fuse Prematurely:** If you replace a blown fuse without addressing the root cause, it's likely to blow again. If the fuse blows immediately after replacing it, this indicates that the short circuit issue hasn't been fully resolved. Investigate and address the underlying problem—such as damaged wiring or a faulty component—before attempting further replacements.

Electrical shorts aren't just annoying—they can lead to more significant problems like drained batteries or malfunctioning systems. Learning to diagnose and repair them not only saves you money but also deepens your understanding of how your car works.

Diagnosing and Fixing a Check Engine Light Issue

The check engine light is one of the most common and dreaded warnings for car owners. It's a general signal that something in your car's system isn't working correctly. While it can indicate a minor issue, like a loose gas cap, it can also signal more significant problems, such as engine misfires or faulty sensors. Tackling this light doesn't have to be intimidating—with the right tools and approach, you can diagnose and resolve the issue yourself.

Tools Needed:

- OBD-II scanner
- Basic hand tools (e.g., screwdriver, wrench)
- Safety gloves

Estimated Time: 30–60 minutes

Instructions:

- **Connect the OBD-II Scanner:** Start by locating the OBD-II port, typically found under the dashboard on the driver's side. Plug the scanner into the port, and turn your car to the "on" position without starting the engine. This powers the scanner and allows it to communicate with your car's computer system.

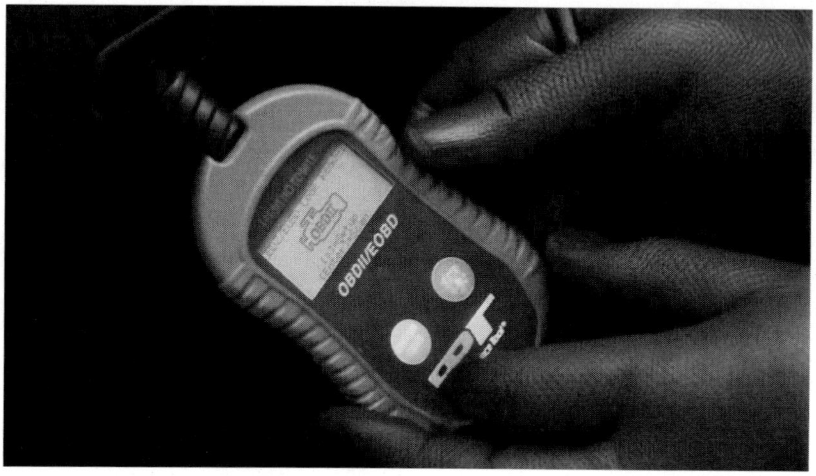

- **Read the Codes:** Follow the scanner's instructions to retrieve diagnostic trouble codes (DTCs). These alphanumeric codes provide clues about what's triggering the light. Write down the codes and consult your car manual or online resources to decode their meaning. Some scanners may offer brief descriptions of the problem, which can save you time.

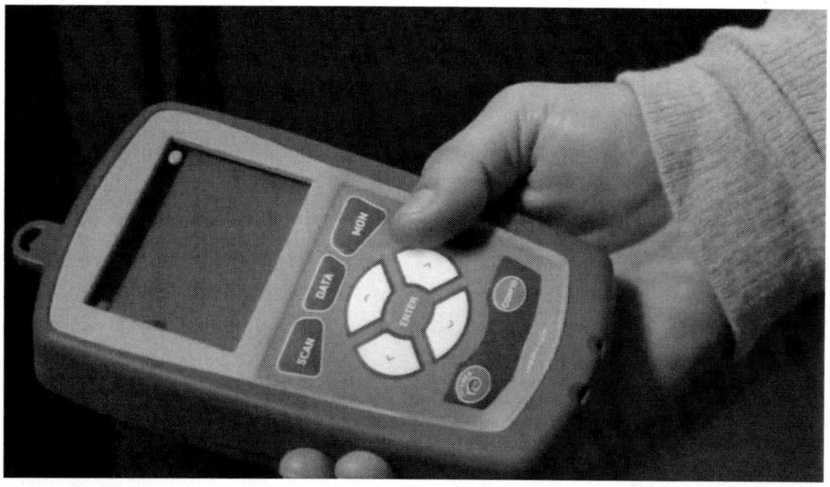

- **Investigate the Issue:** Once you know the general area of the problem, it's time to dig deeper. Common issues include:
  - **Loose gas cap**: A poorly sealed or damaged gas cap can disrupt fuel system pressure. Tighten it and see if the light goes off after a few drives.

- **Faulty oxygen sensor**: If a code points to the oxygen sensor, inspect or replace it.
- **Misfires**: Misfires often result from bad spark plugs or ignition coils.

Use your car manual or online guides for detailed instructions on inspecting specific components.

- **Fix the Problem:** Address the issue based on your findings.
  - **For a loose gas cap**, ensure it's properly tightened.
  - **For faulty spark plugs or sensors**, refer to other sections of this book for guidance on replacement or cleaning procedures.
  - If the issue requires specialized knowledge or advanced tools, consult a professional mechanic to avoid further damage.
- **Clear the Code:** Once the issue is fixed, use your OBD-II scanner to reset the check engine light. Most scanners have a "clear codes" function—follow the device's instructions. After resetting, take your car for a short drive. If the light stays off, you've successfully resolved the problem. If it comes back on, recheck the codes to ensure no additional issues are present.

Safety Precautions:

- **Work in a well-ventilated area**: If you're inspecting components that involve the engine or exhaust system, avoid inhaling harmful fumes.
- **Turn off the engine**: Always switch off the car before handling electrical or mechanical parts.
- **Protect yourself**: Wear gloves to avoid burns or contact with sharp components.

Common Pitfalls and Troubleshooting Tips:

- **Ignoring Simple Issues:** It might sound basic, but always start with the simplest solution. A loose gas cap is one of the most frequent causes of a check engine light and commonly triggers diagnostic trouble codes like P0455 or P0440, which indicate an evaporative emission system leak. Always ensure the gas cap is tightened properly before investigating more complex issues.
- **Clearing Codes Without Fixing the Problem:** Resetting the check engine light won't fix the root cause. If the issue persists, the light will likely come back on. After clearing codes, if the light reappears, take additional steps such as inspecting related components or using the OBD-II scanner again to identify any new or persistent codes. Resolving the underlying issue is crucial for long-term functionality.
- **Overlooking System Interactions:** Many car systems are interconnected. For example, an unresolved oxygen sensor issue can lead to poor fuel efficiency or even damage to the catalytic converter. Address problems promptly to avoid compounding issues.

Handling a check engine light yourself saves time, money, and stress. It's a straightforward way to gain confidence in maintaining your car while ensuring small issues don't grow into costly repairs.

Restoring Your Car's Paint and Preventing Rust

A pristine paint job isn't just about aesthetics—it also protects your car from rust and long-term damage. Rust can spread quickly if left unchecked, weakening your vehicle's structural integrity. The good news? Restoring your car's paint and addressing rust spots is a manageable DIY project that will save you money and keep your car looking great.

Tools Needed:

- Car wash supplies (bucket, sponge, soap)
- Sandpaper (1200–2000 grit)
- Rust remover or converter
- Touch-up paint (color-matched to your car)
- Clear coat
- Polishing compound
- Microfiber cloths

Estimated Time: 2–4 hours

Instructions:

- **Clean the Area:** Start by thoroughly washing your car to remove dirt, grime, and debris. Pay close attention to rusted or chipped areas, as contaminants can interfere with the repair process. Let the area dry completely before moving on.
- **Sand Away Rust:** Before sanding, ensure the surface is clean and completely dry to avoid contaminating the sanded area. Using fine-grit sandpaper (1200–2000 grit), gently sand the rusted spot until you reach bare metal. For stubborn rust, apply a rust remover or converter to neutralize the corrosion, following the product instructions carefully. This step is crucial because paint won't adhere properly to rust, and untreated spots will continue to corrode.

- **Apply Touch-Up Paint:** Once the rust is removed, apply touch-up paint that matches your car's color. You can find color codes in your owner's manual or on a sticker inside the driver's door. Use a small brush or touch-up pen for precision. Apply the paint in thin layers, allowing each coat to dry fully before adding another. This ensures a smooth, even finish.

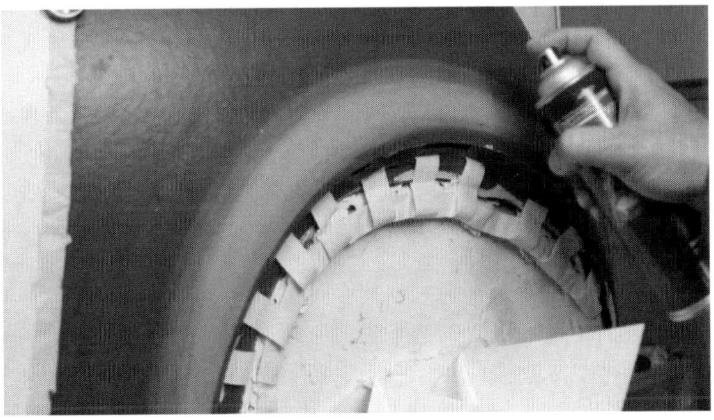

- **Add a Clear Coat:** After the touch-up paint has dried completely, apply a clear coat to seal the repair. The clear coat protects the paint and adds a glossy finish, helping the repaired area blend seamlessly with the rest of the car. Let the clear coat dry thoroughly.

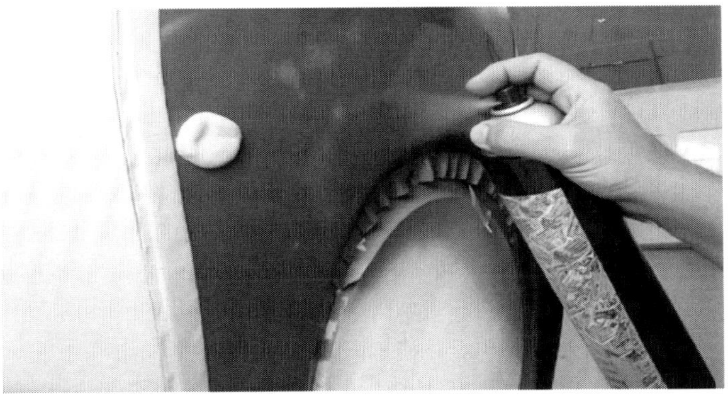

- **Polish the Surface:** Finish the job with a polishing compound to smooth out any imperfections and enhance the shine. Use a microfiber cloth to buff the area in circular motions until it blends seamlessly with the surrounding paint. This final touch gives the repair a professional look.

Safety Precautions:

- **Wear gloves and a mask**: When sanding or using rust removers, protect yourself from chemicals and fine particles.
- **Work in a shaded area**: Direct sunlight can cause paint and clear coat to dry too quickly, leading to uneven application.

Common Pitfalls and Troubleshooting Tips:

- **Skipping Rust Removal:** Never paint over rust. Even a tiny spot can grow and cause the paint to bubble or peel later. Take the time to sand thoroughly and use a rust converter if needed.
- **Rushing the Process:** Each step, from sanding to drying paint and clear coat, requires patience. Skipping drying times can result in uneven layers or a poorly blended repair.
- **Mismatched Paint:** Always double-check the paint color code before purchasing touch-up paint. A mismatched shade will stand out and undo your hard work.

RECAP

This chapter showed just how much you can achieve when it comes to maintaining and repairing your car on your own. From quick, simple wins like replacing wiper blades or a cabin air filter to slightly more advanced tasks like swapping out brake pads or cleaning your throttle body, you've seen how approachable—and satisfying—these DIY projects can be.

Beyond just tackling the repairs, we explored the tools and tips that make the process easier, safer, and more efficient. Whether it's gathering the right equipment or troubleshooting common issues, you now have a practical foundation to take on these tasks with confidence. And let's not forget the savings—every repair you handle yourself is money back in your pocket, not to mention the unique sense of accomplishment that comes with fixing something on your own.

But this chapter wasn't just about saving money; it was about empowerment. You've taken important steps toward becoming a more informed and confident car owner, someone who understands their vehicle and isn't afraid to get their hands dirty. Every DIY project you complete adds to your skills and builds your connection with your car.

That said, it's perfectly okay to recognize when a repair might be outside your skill set or comfort zone. In the next chapter, we'll explore how to decide when it's time to call in a professional. Knowing your limits is just as important as knowing your strengths, and understanding when to hand things over can save you time, stress, and unnecessary expenses. Let's dive into how to make those decisions wisely and keep your car in top shape!

CHAPTER 12

WHEN TO CALL A PROFESSIONAL

SOME REPAIRS ARE BETTER LEFT TO THE PROS —KNOW YOUR LIMITS

A friend of mine once decided to tackle his car's timing belt replacement. He'd watched a few tutorials, gathered the tools, and felt confident he could handle it. Things started smoothly enough, but halfway through, he realized he'd misaligned a critical component. By the time he called a professional for help, his mistake had caused serious engine damage, turning what could have been a manageable repair into an expensive overhaul. It's a perfect example of why knowing your limits with DIY repairs isn't just important—it's essential.

In this chapter, we'll look at when calling a professional is the smartest, safest, and most cost-effective choice. Some repairs are straightforward, but others require advanced tools or expertise that even the most confident DIYer might not have. For example, handling high-voltage systems in electric vehicles or recalibrating advanced driver-assistance features often demands professional-grade equipment and training.

We'll also explore scenarios where sticking to manufacturer specifications or protecting your car's warranty makes professional help the best route. Sometimes, the stakes are simply too high for trial and error—like working on your car's structural integrity or performing complex engine repairs. Knowing when to step back can save you from costly mistakes and unnecessary stress.

Equally important is finding the right mechanic when you do need help. We'll share tips for identifying trustworthy professionals, including how to spot certifications, read reviews, and ask the right questions. Whether it's choosing the best shop or setting clear expectations for a repair, knowing what to look for will keep you in control.

By the end of this chapter, you'll have a clearer sense of when to tackle a repair yourself and when to let a pro handle it. Together, we'll ensure your car stays in great shape while keeping your confidence—and budget—intact.

RECOGNIZING THE LIMITS OF DIY REPAIRS

DIY car repair can be incredibly rewarding—you save money, learn new skills, and feel the satisfaction of fixing something with your own hands. But it's important to know where to draw the line. Some repairs are best left to professionals, not just for your safety but also for the long-term health of your vehicle. So, how do you know when to step back? Let's break it down.

The Complexity of the Problem

Today's cars are more like rolling computers than machines. Modern vehicles are packed with intricate systems that communicate through a network of sensors and control modules. Diagnosing engine malfunctions, electrical issues, or problems with advanced driver-assistance systems (ADAS) isn't always straightforward. These systems often require highly specialized tools and training to pinpoint faults accurately. If you're dealing with multiple interconnected components or errors that don't seem clear-cut, it's probably time to consult a professional.

Specialized Tools and Equipment

Not all repairs can be done with the tools in your garage. Some jobs demand precision equipment, like diagnostic scanners, alignment machines, or tools for recalibrating sensors. For example, recharging an air conditioning system or fixing suspension alignment requires instruments that professionals use daily—but that the average DIYer rarely owns. Not only are these tools expensive, but they also require experience to operate properly. Skipping this step or improvising with the wrong tools could make the problem worse, leading to more costly fixes later.

Manufacturer Specifications

When working on your car, you need to follow the manufacturer's guidelines to the letter. These specifications—like torque settings, part compatibility, and calibration procedures—are there for a reason. Ignoring them, even unintentionally, can lead to significant damage or even void your vehicle's warranty. For example, tightening engine components without the proper torque can cause long-term engine failure. If you don't have access to detailed repair manuals or the right technical data, it's best to let a professional handle it.

Uncertainty or Lack of Confidence

Let's face it: If a repair feels overwhelming or you're second-guessing yourself at every step, it's a sign to take a step back. Doubt can lead to costly mistakes. For instance, installing a timing belt incorrectly might seem minor, but it can cause catastrophic damage to the engine. If you're watching tutorials on repeat or feeling unsure about the repair process, it's okay to hand the job over to a pro. A professional mechanic will not only have the tools and knowledge but will also ensure the work is done correctly the first time.

Know Your Limits—And Be Okay with Them

Knowing when a repair is beyond your skill level isn't admitting defeat—it's making a smart decision for both you and your car. Tackling manageable tasks, like changing the oil, swapping filters, or replacing brake pads, is a great way to gain confidence and experience. But when you're dealing with complex systems, specialized tools, or uncertainty, seeking professional help can save you time, stress, and money in the long run.

Ultimately, being a smart DIYer is about balance: doing what you can confidently handle and recognizing when to let the experts take over. Your car—and your wallet—will thank you.

SAFETY CONCERNS FOR ADVANCED REPAIRS

When it comes to car repairs, your safety should always be the top priority. While basic tasks like oil changes or replacing filters can be done with minimal risk, some repairs are inherently dangerous if not handled with the right tools, training, and precautions. Let's dive into the situations where safety concerns make professional help the smarter choice.

Working with High-Voltage Systems

Electric and hybrid vehicles are becoming more common, but they come with their own set of challenges. The high-voltage batteries in these cars can deliver a fatal shock if mishandled. Unless you're specifically trained to work on electric systems, it's best to steer clear. These repairs often require specialized tools, like insulated gloves and high-voltage meters, along with strict safety protocols that go far beyond standard DIY practices.

Brake and Steering Systems

Your brakes and steering aren't just essential—they're life-saving systems. While tasks like replacing brake pads or topping off power steering fluid are manageable, repairs that go deeper into the hydraulic or electronic systems demand extreme precision. A single mistake could compromise your ability to stop or steer safely. For instance, working on anti-lock braking systems (ABS) requires careful calibration and handling of complex electronic components. It's one area where guessing simply isn't an option.

Structural Repairs

Damage from accidents isn't always about dents and scratches—sometimes it's structural. A bent frame, for example, can affect the car's alignment, safety, and overall performance. Fixing this kind of damage requires specialized equipment like frame alignment machines, which DIY methods can't replicate. Professional body shops have the tools and expertise to restore structural integrity, ensuring your car is safe to drive.

Handling Hazardous Materials

Some repairs involve hazardous materials that can harm you or the environment if mishandled. For instance, refrigerant in air conditioning systems must be carefully recovered and replaced using specialized recovery machines. Similarly, working with gasoline or fuel lines can present serious fire risks. If a repair requires handling toxic, flammable, or pressurized substances, it's safer to let a professional take over.

Heavy Lifting

Repairs that involve lifting the car off the ground—like replacing a transmission, differential, or exhaust system—require sturdy, heavy-duty equipment. While jack stands and hydraulic jacks can work for smaller jobs, inadequate tools or improper lifting techniques can result in catastrophic accidents. Professional repair shops use high-capacity lifts that make these tasks safer and more efficient, minimizing the risk of injury.

Know When to Call in the Pros

While DIY car care can save you time and money, it's important to recognize when a repair crosses into risky territory. Complex systems, hazardous materials, or heavy equipment aren't things to take lightly. Knowing your limits and trusting professionals for advanced repairs keeps you safe—and ensures your car is properly fixed. After all, your safety is worth far more than the cost of any repair.

HOW TO FIND A TRUSTWORTHY MECHANIC

Knowing when to call in the pros is one thing, but is finding a mechanic you can actually trust? That's where things can get tricky. A bad experience can leave you feeling overcharged, frustrated, and no better off than when you started. So how do you separate the reliable shops from the ones just looking to pad their bill? Here's how you can choose wisely.

Research and Reviews

Start by doing a little homework. Websites like Yelp, Google Reviews, and automotive forums are gold mines for feedback. Look for consistent praise about honesty, fair pricing, and quality workmanship. One glowing review is great, but if you see a pattern of happy customers, that's even better. On the flip side, avoid shops that get frequent complaints about pushing unnecessary services or charging surprise fees.

Certifications and Credentials

Next, check for credentials. A trustworthy mechanic will often have certifications like ASE (Automotive Service Excellence). This badge shows they've undergone rigorous training and testing to prove their skills. Some mechanics even hold certifications from specific car manufacturers, which means they have specialized knowledge about your particular make and model. It's a good sign they know what they're doing.

Ask for Recommendations

Sometimes, the best resource is right in front of you. Ask your friends, family, or coworkers who they trust with their cars. A personal recommendation can give you insights that online reviews can't—like how friendly the staff is, how quickly they work, or how transparent they are with pricing. Don't underestimate community groups or social media platforms, either. Locals often share their go-to shops and experiences, which can point you in the right direction.

Check for Transparency

A good mechanic won't just fix your car—they'll take the time to explain what's wrong, what needs to be done, and why. They'll provide a detailed estimate and answer your questions without hesitation. If a mechanic seems evasive, dismissive, or pushy, take it as a red flag. A transparent approach builds trust and ensures you're on the same page before any work begins.

Look for a Warranty

Reputable mechanics stand by their work. Look for shops that offer warranties on both parts and labor—this shows they're confident in the quality of their repairs. For instance, a one-year warranty on repairs is pretty standard and gives you peace of mind knowing they'll make things right if something goes wrong.

COMMUNICATING EFFECTIVELY WITH YOUR MECHANIC

Clear and open communication with your mechanic can make all the difference in ensuring your car gets the service it needs—without unnecessary surprises. Here are some tips to help you build a productive and stress-free interaction.

Describe the Issue Clearly

Start by sharing as much detail as possible about what's happening with your car. When does the problem occur? Are there any specific sounds, smells, or changes in performance? For example, saying, "I hear a squealing noise when I brake, and it stops when I let off the pedal," gives the mechanic a great starting point. The clearer you are, the easier it is for them to diagnose the issue accurately.

Ask Questions

Don't shy away from asking questions. A good mechanic will welcome them and explain everything in terms you can understand. Ask about the repair process, what parts are being used, and how much it's going to cost. Understanding what's being done not only helps you feel confident but also ensures you're not caught off guard when the bill comes.

Set Expectations

If you're working with a budget, be upfront about it. Let the mechanic know you want to be contacted before they proceed with any additional work outside the original estimate. This way, you stay in control of the costs and avoid any unexpected surprises. A simple "Please call me if anything comes up beyond what we've discussed" goes a long way in setting clear expectations.

Request Old Parts

Asking to see the parts that were replaced is a great habit to get into. Not only does it give you peace of mind that the work was done as promised, but it also helps you learn more about your car. For example, seeing a worn timing belt or frayed brake pad makes it clear why those replacements were necessary.

Keep Records

Finally, always keep a record of the work done on your car. Detailed service records help you stay on top of future maintenance and can save you time diagnosing recurring problems. Plus, if you ever decide to sell your car, a complete service history shows prospective buyers that the vehicle has been well cared for.

ADDITIONAL TIPS FOR MAINTAINING A BALANCED APPROACH

Taking care of your car doesn't mean you have to choose between doing everything yourself or relying entirely on professionals. A balanced approach often works best, saving you time, money, and stress. Here's how to strike that balance.

Weighing Costs and Benefits

DIY repairs can be a great way to cut costs, but not every job is worth tackling on your own. Some services—like a comprehensive diagnostic check—are worth the investment. A professional can pinpoint issues you might miss, saving you from wasted time and potential mistakes. Before deciding whether to handle a repair yourself, weigh the cost of professional help against the risk of making the problem worse.

Building a Relationship with Your Mechanic

When you do need professional assistance, having a go-to mechanic you trust is invaluable. Once you find someone reliable, stick with them. Over time, they'll become familiar with your car's history, which often leads to quicker diagnoses and more tailored advice. Plus, loyalty has its perks—you might get discounts, priority scheduling, or honest guidance about what repairs can wait. A good relationship with your mechanic benefits you in the long run.

Staying Educated

Even if you're leaving complex repairs to the professionals, staying informed about your car's systems gives you confidence and control. Understanding the basics—like what the transmission does or how to check your fluids—makes you a better car owner. You'll communicate more effectively with mechanics and be less likely to approve unnecessary repairs. Knowledge is power when it comes to car maintenance.

Combining DIY and Professional Repairs

The sweet spot is often a mix of DIY and professional work. Handle the small, manageable tasks yourself—like changing the oil, swapping out an air filter, or replacing windshield wipers. For more advanced repairs, like fixing suspension components or diagnosing complex electrical problems, call in the professionals. This balanced approach lets you maximize savings while minimizing risks.

RECAP

This chapter was all about recognizing when it's smarter—and safer—to let a professional take over. While DIY repairs can save you money and give you a sense of accomplishment, knowing your limits is just as important. We explored repairs that require specialized tools and advanced expertise or come with safety risks, like working with high-voltage systems, handling hazardous materials, or aligning complex sensors. Tackling these jobs on your own can quickly turn into a costly or even dangerous mistake, which is why they're best left to the experts.

Finding the right mechanic can feel daunting, but it doesn't have to be. We discussed how to read reviews, check for certifications, and ask the right questions to ensure you're working with someone trustworthy. A reliable mechanic isn't just a one-time help—they're a partner in keeping your car running smoothly. Building a good relationship with your mechanic can save you from unnecessary upsells, unexpected charges, and countless headaches.

That said, calling in a professional doesn't mean abandoning DIY entirely. We emphasized the value of a balanced approach: handling simpler tasks yourself, like changing filters or swapping out wiper blades, while knowing when to step back for the bigger jobs. This way, you save money, build your confidence, and keep your car in peak condition without taking unnecessary risks.

CONCLUSION

As we wrap up this book, let's pause to reflect on where we started and where you've arrived. When you first picked up this book, maybe you felt unsure about car maintenance—how it works, where to start, or whether you could handle it yourself. Perhaps you were frustrated by the money you'd spent on repairs that seemed mysterious or avoidable. Whatever brought you here, I hope you're leaving with a clear sense of direction and confidence.

This book wasn't written to turn you into a professional mechanic. It's not about tackling every issue under the hood or pretending that car care is always simple. Instead, it's about giving you the tools, knowledge, and mindset to approach your car with greater understanding and control. You've gained the skills to handle routine maintenance, make informed decisions, and save money without feeling at the mercy of a repair shop.

We began by breaking things down to the basics: understanding your car's key systems and what they need to keep running smoothly. The engine, brakes, suspension, and other systems might have seemed complex at first, but now you know they're manageable with the right approach. Recognizing common signs of wear and knowing when to act is a big step—one that puts you ahead of most car owners.

Next, we focused on getting you set up with the essentials. With a small investment in affordable tools, you can handle a surprising number of tasks on your own. You don't need a workshop or a big budget to start; just a few basics like wrenches, screwdrivers, and a diagnostic scanner are enough to get you going. Having those tools—and knowing how to use them—puts you in a position to solve problems before they escalate.

Preventative maintenance came next. This is where the real savings happen. By adopting simple habits like checking fluid levels, inspecting your tires, and keeping an eye on your battery, you're preventing problems before they start. These habits might seem small, but they add up. A few minutes here and there can save you thousands of dollars over the life of your car—not to mention the stress of an unexpected breakdown.

We also covered common repairs. Whether it's replacing brake pads, dealing with a dashboard warning light, or swapping out a headlight bulb, these are tasks you now have the know-how to tackle. They're not as intimidating as they seem once you break them into manageable steps. And even if things don't go perfectly the first time, that's okay. Every time you take on a new project, you're learning and building confidence for the next one.

Of course, knowing your limits is just as important as knowing what you can do. There are some repairs that require specialized tools or professional expertise. Being able to recognize those situations and knowing when to call in a mechanic isn't a failure—it's part of being a responsible car owner. And now, when you do need professional help, you'll know how to ask the right questions and avoid unnecessary expenses.

Reflecting on this journey, it's clear that car care is about more than just fixing things. It's about taking ownership. It's about shifting from a mindset of uncertainty to one of control. It's about turning what might have once felt frustrating into something empowering. You've gone from wondering if you can handle your car's needs to knowing that, in many cases, you absolutely can.

So, what's next? Maybe it's time to try something simple, like checking your oil or replacing your wiper blades. Or perhaps you're ready to tackle a bigger project, like changing your brake pads. Wherever you start, remember that progress is the goal. Car maintenance isn't a one-time task—it's an ongoing process. The more you do, the more you'll learn. The more you learn, the more confident you'll feel.

As you move forward, I encourage you to think about car care in a broader sense. Beyond saving money and keeping your vehicle in good shape, there's an opportunity to be more resourceful and even more environmentally conscious. Recycling old parts, reducing waste, and helping others learn basic car care are all ways to make a difference. Sharing what you've learned with a friend or family member who feels overwhelmed by their car can be incredibly rewarding.

Above all, car maintenance is about building trust—trust between you and your vehicle. When you understand what your car needs and how to meet those needs, you'll feel more in control. You'll know that when you start the engine, your car is ready to take you wherever you need to go. That's what this is really about: independence, reliability, and the satisfaction of knowing you're capable of handling what comes your way.

Thank you for taking the time to read this book. I know there are countless ways you could have spent your time, and I appreciate that you chose to spend some of it here. My hope is that the knowledge and skills you've gained will make your life just a little easier—and maybe even a little more enjoyable.

As you close this book and move on to your next step, remember that you don't need to have all the answers right away. Start with what you know, and keep building from there. Every check you perform, every repair you make, and every dollar you save is a testament to your growth as a car owner.

The road ahead is yours. You're ready—go make the most of it.

Made in United States
North Haven, CT
10 November 2026